Microsoft

D0796854

Office

Excel 2007

A Beginners Guide

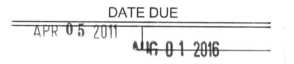
By WR Mills

AuthorHouse™
1663 Liberty Drive
Bloomington, IN 47403
www.authorhouse.com
Phone: 1-800-839-8640

First published by AuthorHouse 1/6/2010

IBSN: 978-1-4490-3233-3 (e)
ISBN: 978-1-4490-3232-6 (sc)

Library of Congress Control Number: 2009913627

Printed in the United States of America
Bloomington, Indiana

This book is printed on acid-free paper.

Microsoft

Office

Excel 2007

A Beginners Guide

A training book for Microsoft Excel 2007

By WR Mills

About the Author

Bill has a background in electronics and technology. He started writing software in 1982 and has expanded his programming skills to include C, C++, and Visual Basic. Bill also designs web sites. He designed a computer based telephone system for the hotel/motel market and wrote the entire operating system himself.

In 2007 he started teaching computer training classes and seems to have a knack for explaining things in a simple way that the average user can understand.

Bill is self-employed and lives in Branson Missouri with his wife Rose. They have three children, two sons and a daughter.

Preface

In 2007 I started teaching computer training classes. I was shocked at how much trouble the students had trying to understand the textbooks. I spent all of my time explaining what the textbook was trying to get across to the reader. It wasn't until I started getting ready for teaching the Microsoft Office 2007 series of classes that I finally gave up and started writing the textbooks myself.

These books are easy to understand and have step by step, easy to follow, directions. These books are not designed for the computer geek; they are designed for the normal everyday user.

It seems I have a knack for explaining things in a simple way that the average user can understand. I hope this book will be of help to you.

William R. Mills

Foreword

Dear Bill, I wanted to write a note of appreciation to you for your books: Microsoft Office Word 2007, Microsoft Office Excel 2007, and Microsoft Office PowerPoint 2007. I've used them all and found each one to be easy to read and very user friendly. If anyone needs to learn one of the 3 programs, but is even a little intimated, I strongly suggest they try one of your books. It's almost as good as taking a class with you as the instructor. If I didn't understand a step, I just went back to the pervious step and tried it again --- and it always worked!! There's just enough humor in the text to keep the reading interesting; never dull, but fun and light. Just what a beginner needs. Again, Bill, I thank you for creating these books that make learning something I needed to learn fun and easy. Sincerely, Cyndy O

Important Notice:

There will be times during this book that you be asked to open a specific file for the lesson. These files can be downloaded from the EZ 2 Understand Computer Books web site.

Open your internet browser (probably Internet Explorer) and go to www.ez2understandcomputerbooks.com. Click your mouse on the <u>Lesson Files</u> link toward the top. This will take you to the page where you can download the files needed. There are directions on the page to help you with the download. They are repeated below for your convenience.

To download the files follow the following steps:

1)   Right-click your mouse on the file(s) you want. These are zipped files and contain the lessons that you will need for each book.

2)   Select the "Save Target As" choice. Make sure the download is pointed to a place on your hard drive where you can find it, such as My Documents.

3)   Click the Save button

4)   The files are zipped files and will need to be extracted to access the contents. To extract the files right-click on the file and choose "Extract All". Make a note as to where the files are located, so you will be able to find them when you need them.

Table of Contents

Chapter One   Microsoft Excel - The Basics

Microsoft Excel is a powerful spreadsheet application. Its power comes from being able to do very complicated and exact mathematical (numerical) calculations. Do you want to know what is even better? Excel 2007 is easy to use, as you will see. All you have to do is enter the data and tell Excel to perform the needed calculations. This can be used for financial reports, keeping statistics, even setting up your family budget.

The first thing you are going to notice is that this version of Excel looks different than any other version of Excel. This is because of the new user interface. You might ask why this is better than the version you are use to. Do you remember searching through a series of menus and submenus to find a command? That is all a thing of the past. Excel 2007 has the Ribbon. Wow! Are you excited yet? Is the Ribbon scary? Probably. Is it intimidating? More than likely. Is it better and easier to use? Yes definitely. The Ribbon is based more on how people actually use their computer.

The Ribbon is divided into Task Orientated Tabs. Each tab has groups of related commands. Everything you need is right at your fingertips. You will not have to search through menus and submenus until you want to pull your hair out, trying to find a command.

Do you remember the old days when you would copy a large database into an Excel spreadsheet only to find out that it couldn't hold all of the data? Well, I remember. In Excel 2007 you can have up to 1,048,576 rows and 16,384 columns. Life is getting better. Excel can now do faster calculations, since it uses multiple processors and multiple threads.

Enough of this, let's get started using Excel 2007.

Lesson 1 – 1   Starting Excel

The first thing you have to do is have your computer on and the desktop showing. I know, I didn't have to say that, but I did, too late to take it back now. The Excel program is located in the Microsoft Office folder.

Click on the Start Button

Select All Programs

Move the mouse to Microsoft Office and click on Microsoft Excel 2007 from the menu that slides out to the right side

In a few moments, Excel will appear on your screen. It should look similar to Figure 1-1. The options at the top may look slightly different on your screen; it depends on the available screen width of your monitor.

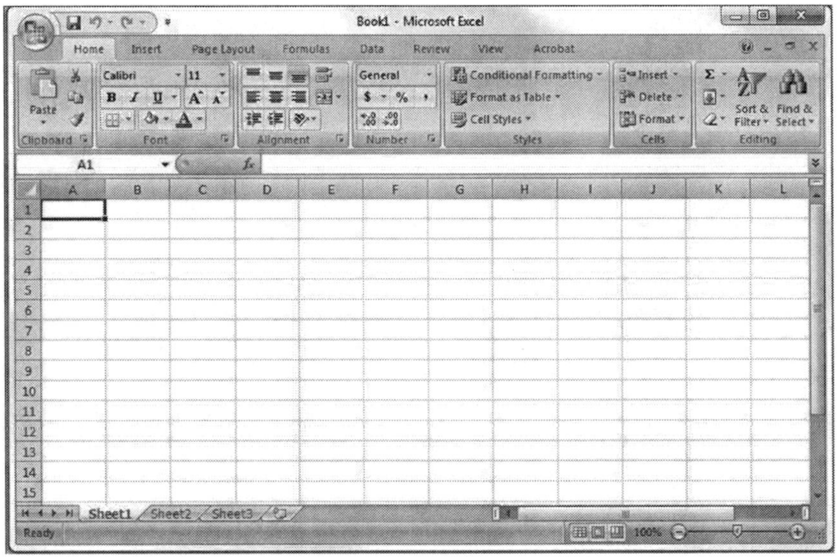

Figure 1-1

Excel 2007 is now ready for you to start entering data. From here you can enter data, format the data, and perform calculations.

Before we start doing any of these things, let's get use to the screen.

Lesson 1 – 2   Understanding the Excel Screen

The moment you start Excel 2007 you will notice some major changes. Microsoft completely redesigned the interface. Microsoft pretty much went back to the drawing board to design the way you use Excel. Now how it works is based on how most people actually use the program. Figure 1-2 shows the Excel screen.

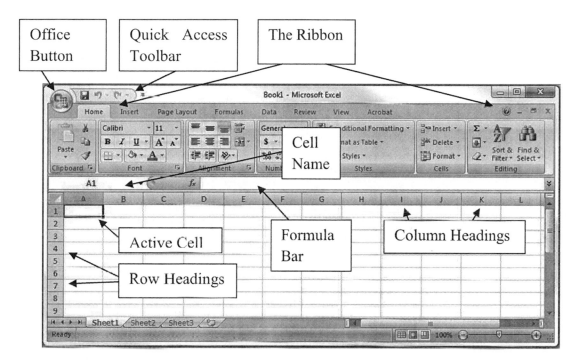

Figure 1-2

Figure 1-3 shows a close-up view of the Office Button. It is actually located on the end of the Quick Access Toolbar. We will discuss the Office Button more in the next lesson.

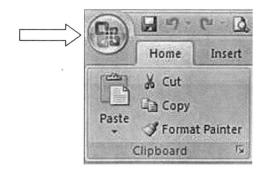

Figure 1-3

3

Figure 1-4 shows the Quick Access Toolbar. We will discuss it in lesson 1-5.

Figure 1-4

The Ribbon is new and we will be using it extensively during almost every lesson in this book. Lesson 1-4 is devoted to explaining the various sections (Tabs) of the Ribbon.

Figure 1-5 shows the Cell Name box. We will be using this as we cover the upcoming lessons. This shows the name of the active cell, the one that is surrounded by a dark line and ready for you to enter data into.

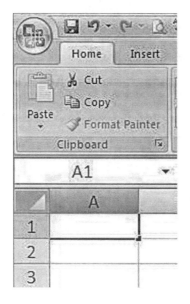

Figure 1-5

Figure 1-6 shows the Formula Bar. This will show any data that is entered into the active cell. This is also where you can edit any data or formula. We will be using the Formula Bar quite a lot as we continue. The Formula Bar actually starts at the end of the Cell Name box. The part with the X and the check mark are part of the Formula Bar.

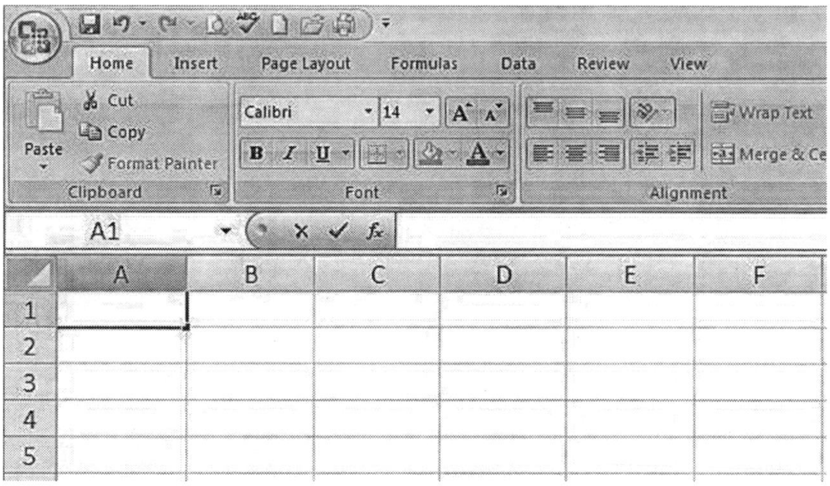

Figure 1-6

The columns are identified by the letters at the top of the column. The far left is column A, and next to it is column B. This continues on until you do not need any additional columns. If you need columns beyond the letter Z the next column would be AA,

The rows are identified by the numbers on the left side of the row. The first row is number 1 and the second row is number 2. This sequence will continue down until you don't need any other rows.

A cell is referenced by the column and the row it is in. This is used in referencing data and is used extensively in formulas.

Lesson 1-3   The Office Button

The Office Button, for the most part, has taken the place of the old File section of the menu bar. As you can see the menu bar does not exist in this version of Excel. In this lesson we will examine the Office Button and see just how it works.

Using your mouse, click on the Office Button

When you click on the Office Button, a menu will drop down giving you several choices of what you are able to do. This is shown in Figure 1-7.

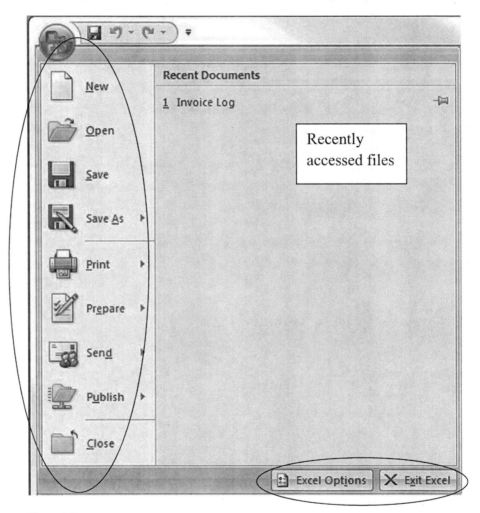

Figure 1-7

On the left side you will notice that many of the choices were the same as when you clicked on the file button of the older style menu bar. You can start a new worksheet or open an existing worksheet. You also have the Save and Save As choices. You may notice that the drop-down choices are divided into two sections. The most popular choices are put at the top. The lesser used options are placed toward the bottom.

If you needed to print the workbook, you would find the printer options under the print choice. The Prepare section is where you would look at the properties of the workbook as well as encrypt it so no one could open it or edit it without a password.

The Send option is where you could send the workbook as an e-mail or a fax.

The Publish section is where you would share the workbook with others or create a new site for the workbook.

In the Business Contact Manager, you can link this workbook to the communications history of a business record.

On the right side are several of the most recent workbooks that you have opened. To open one of these workbooks you simply click on it with the mouse.

The last choice on the left is where you click to close a workbook.

On the bottom right you will see that you can also exit Excel from this part of the menu.

Also on the bottom right is a button to access the Excel Options. The first screen of the available options is shown starting in Figure 1-8.

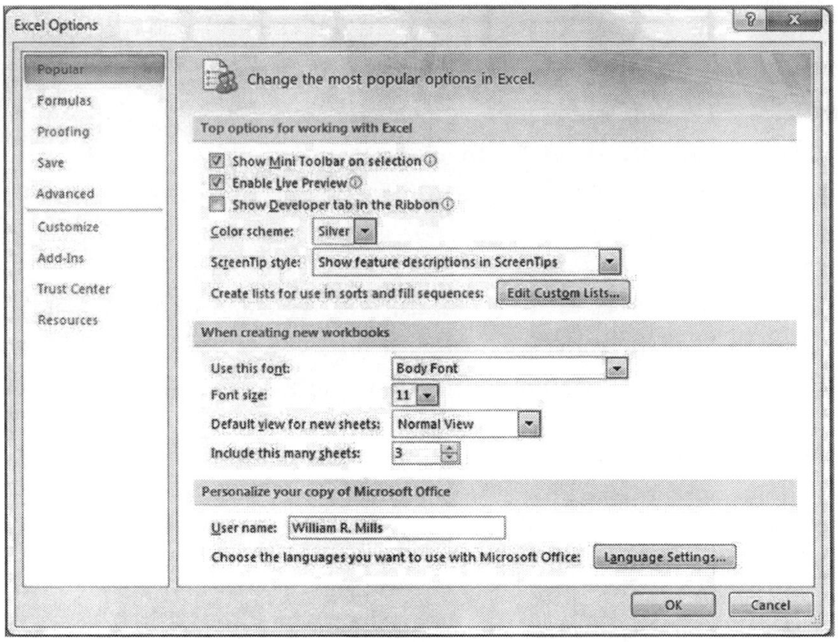

Figure 1-8

From this screen you can change the most popular options, such as changing the font and the font size as well as how many sheets will be in the workbook.

In the Formulas section, shown in Figure 1-9, you can work with the formulas and also with the error checking.

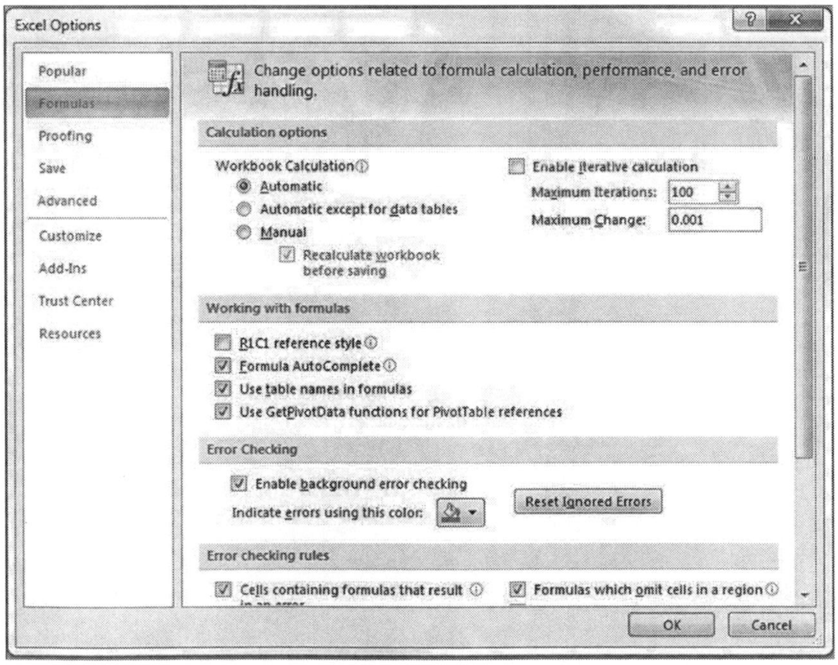

Figure 1-9

In the Proofing section, shown in Figure 1-10, you can change how Excel corrects formulas and text as you type. You can also change how the spell check is working.

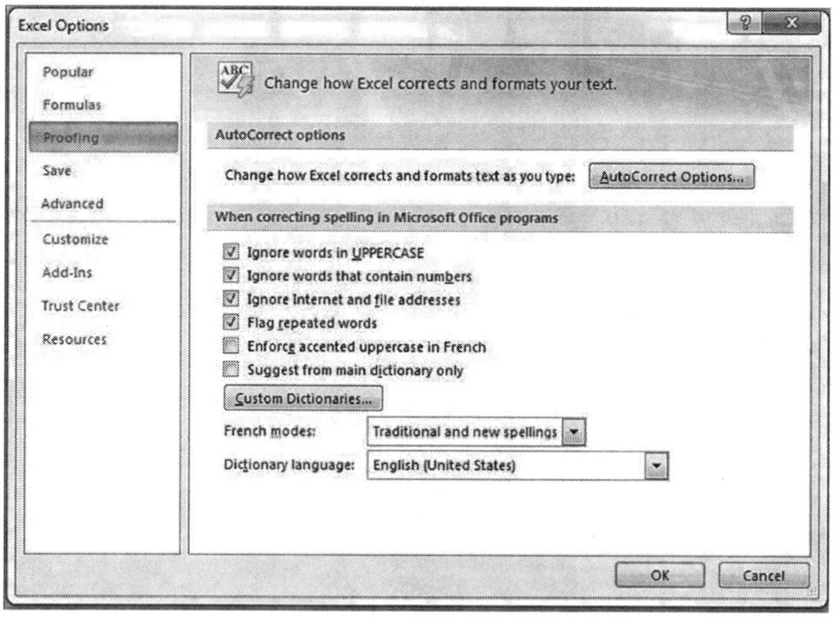

Figure 1-10

In the Save section, you can customize how workbooks are saved. This screen is shown in Figure 1-11.

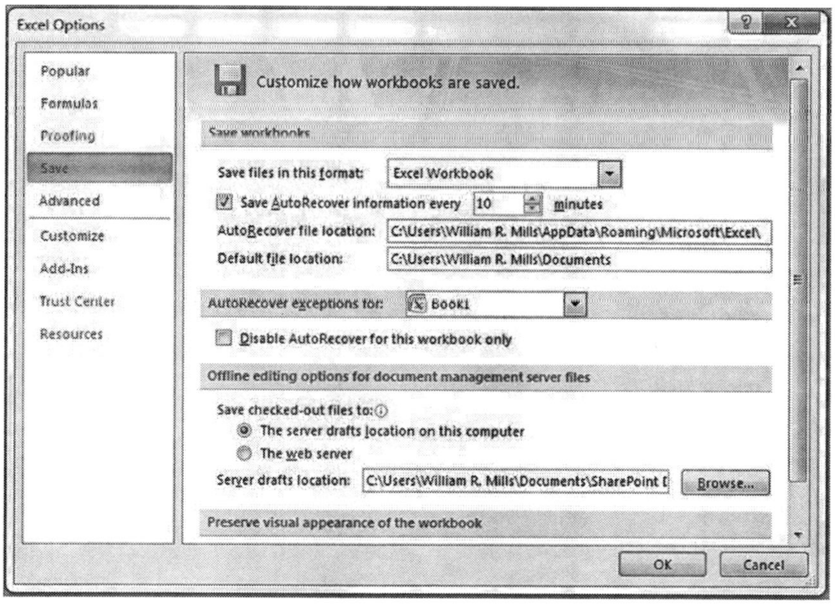

Figure 1-11

In the advanced section you can make changes to such things as: are the fill handles enabled and if you can edit directly inside a cell. You can also change where the new active cell will be when you press Enter on the keyboard. This screen is shown in Figure 1-12

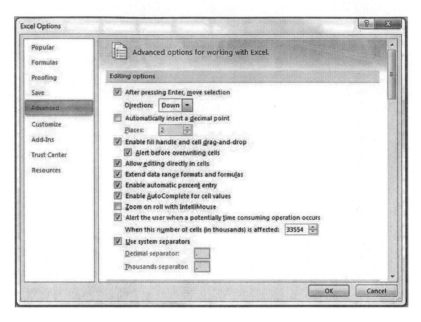

Figure 1-12

The Customize section allows you to add and remove icons from the Quick Access Toolbar. You can also move the toolbar to below the Ribbon instead of having it above the Ribbon. See Figure 1-13 for the Customize screen.

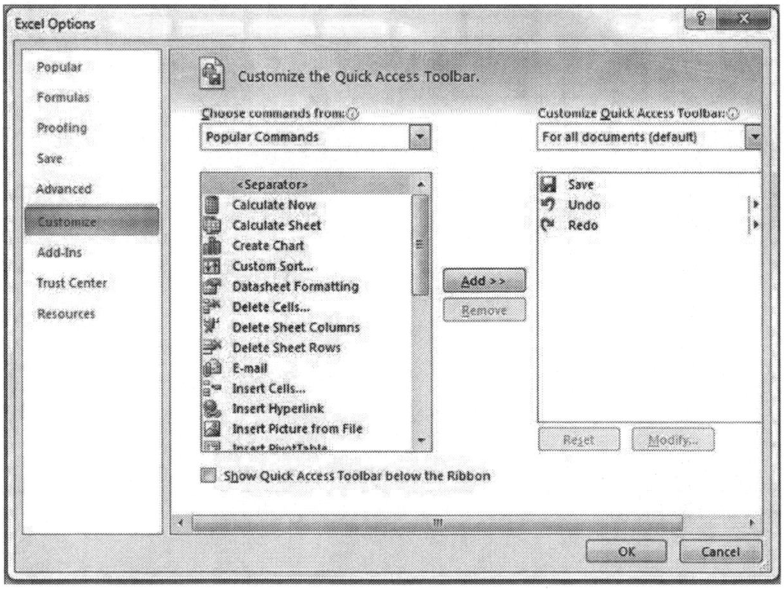

Figure 1-13

This section will be covered in the Quick Access Toolbar lesson.

The add-ins section shows the "extra" things that have been added to help Excel work better. This is shown in Figure 1-14.

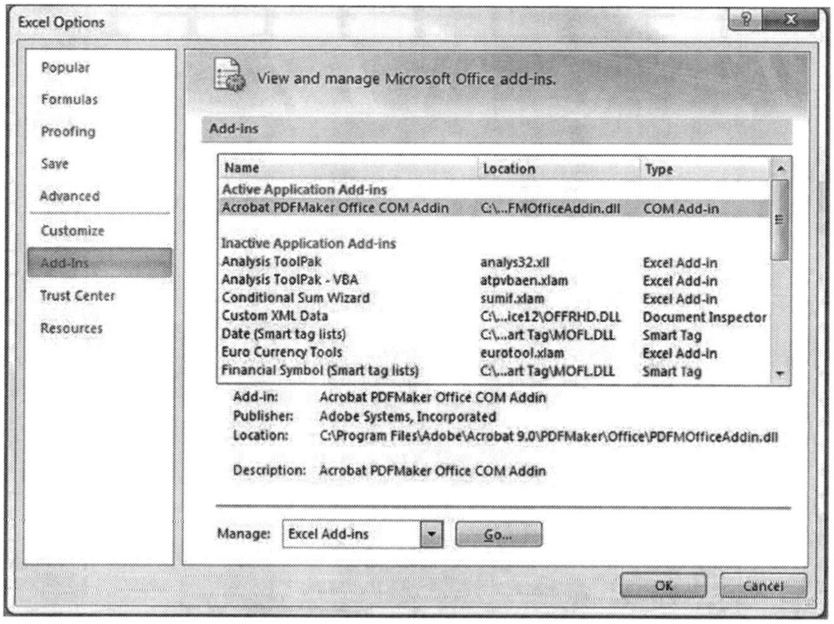

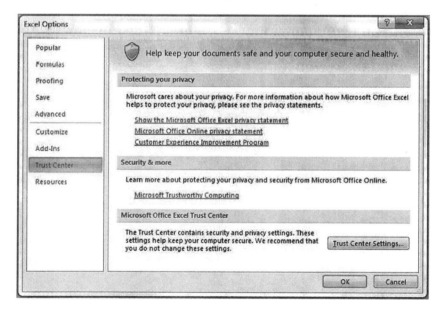

Figure 1-14

The Trust Center contains security and privacy settings. Microsoft recommends that you do not change these settings. See Figure 1-15 for the Trust Center.

Figure 1-15

The last section of the Excel Options is the Resource Center. From here you can get updates, contact Microsoft, etc. this screen is shown in Figure 1-16.

Figure 1-16

Lesson 1 – 4   The Ribbon – An Overview

The Ribbon has been designed to offer easy access to the commands that you (the user) use most often. You no longer have to search for a command embedded in a series of menus and submenus. The Ribbon has a series of Tabs and each tab is divided into several groups of related commands. Figure 1-17 shows the Ribbon across the top of the Excel program.

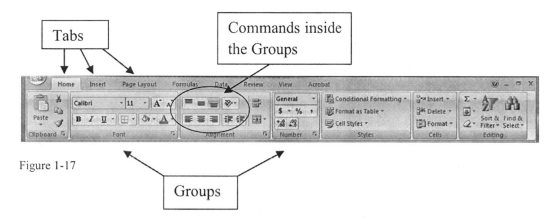

Figure 1-17

There are three major components to the Ribbon.

Tabs:

There are seven basic tabs across the top.

The Home Tab contains the commands that you use most often.

The Insert Tab contains all of the objects that can be inserted into a workbook.

The Page Layout Tab contains the choices for how each page will look.

The Formulas Tab contains the different formulas, names, functions, and lookup functions.

The Data Tab contains such things as getting external data, data tools, and filters available.

The Review Tab has things related to proofing, protecting, and comments.

The View Tab allows you to change to the different views that are available.

Groups:

Each Tab has several Groups that show related item together.

Look at the Home Tab to see an example of the related Groups.

The Home Tab has the following Groups: Clipboard, Font, Alignment, Number, Styles, Cells, and Editing.

Commands:

A Command is a button, a box to enter information, or a menu.

The Clipboard Group, for example, has the following commands in it: Cut, Copy, Paste, and Format Painter.

When you first glance at a group, you may not see a command that was available from the menus of the previous versions of Excel. If this is the case you need not worry. Some Groups have a small box with an arrow in the lower right side of the Group. See figure 1-18 for a view of a group with this arrow.

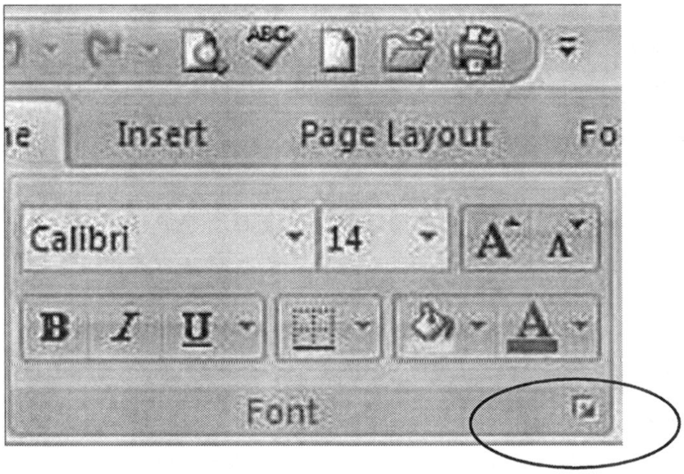

Figure 1-18

This small arrow is called the Dialog Box Launcher. If you click on it, you will see more options related to that Group. These options will usually appear in the form of a Dialog Box. You will probably recognize the dialog box from previous versions of Excel. These options may also appear in the form of a task pane. Figure 1-19 shows the Font Dialog Box.

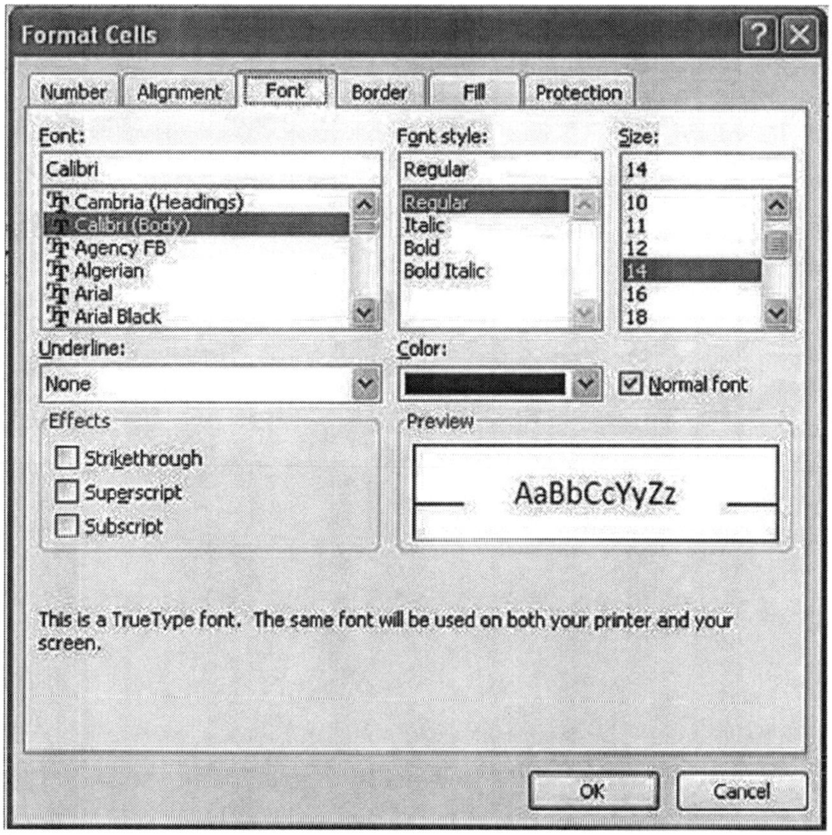

Figure 1-19

Speaking of previous versions, if you are wondering whether you can get the look and feel of the older versions of Excel back, the answer is simple, no you can't.

The good news is that after playing with and using the Ribbon, you will probably like it even better. It really does make working with the spreadsheet easier. The Ribbon will be used extensively and each tab covered in more detail later as we go through this book.

Lesson 1 – 5   The Quick Access Toolbar

The Ribbon, as you will find out, is wonderful, but what if you want some commands to always be right at your fingertips without having to go from one tab to another? Microsoft gave us a toolbar for just that purpose. This toolbar is called the Quick Access Toolbar and is located just above, or below, and to the left end of the Ribbon. The Quick Access Toolbar is shown in Figure 1-20.

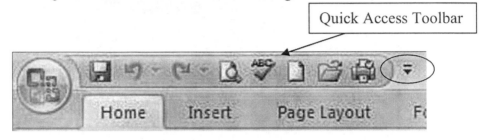

Figure 1-20

The Quick Access Toolbar contains such things as the Save button, the Undo and Redo button, the Quick Print button, and Spell Check button. These are things that you normally use over and over and you will want them available all of the time.

There is even more good news, if you want to add an item to the toolbar, the process is very simple. At the right end of the toolbar is an arrow pointing downwards. If you click on this arrow, a new drop down menu will come onto the screen, as shown in Figure 1-21.

From this menu you can choose from the standard choices or you can customize the toolbar to suit your needs by clicking on the More Commands choice.

You can also choose to show the tool bar below the Ribbon instead of above it. I have my computer set to show the Quick Access Toolbar below the Ribbon, probably because the toolbars were always below the menu bar in the older versions.

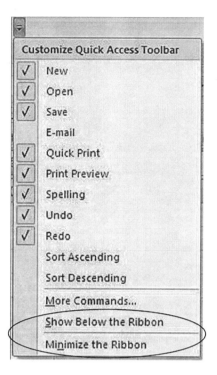

Figure 1-21

If you want to add an item from the standard choices all you have to do is click on the item you want to add. The drop down menu will disappear and the new item will be added to the toolbar.

Add the items that are checked in Figure 1-21 to your Quick Access Toolbar

If the option you want to add is not listed in the standard choices, all of the available options are listed under the More Commands.

Figure 1-22 shows the Excel Options Dialog box that will come to the screen if you choose the More Commands option.

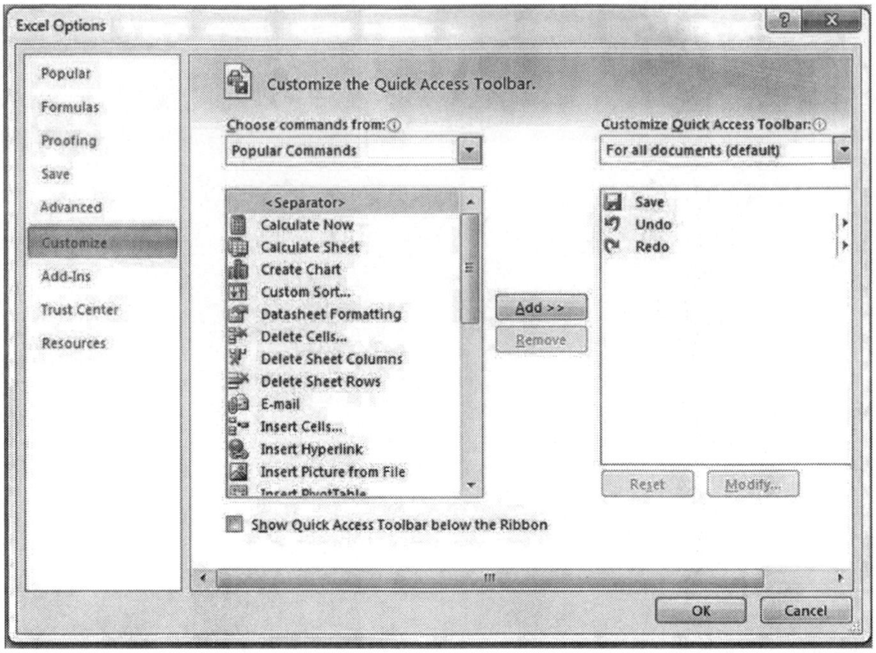

Figure 1-22

If you wish to add an item to the Quick Access Toolbar, all you need to do is click on the option on the left and then click the Add button in the center. When you are finished adding items, click the OK button to place them in the toolbar. You will probably find that there are several things that you will use over and over with every workbook and you will want to place them in the Quick Access Toolbar just because this will save you so much time.

Lesson 1-6   Using the Keyboard

What about all of you people who prefer to use the Keyboard over the mouse? I have not forgotten about you, and this lesson is just for you. People who prefer the keyboard over the mouse often started way back with DOS. Back then, in the olden days, you had to use the keyboard to do everything. That is a hard habit to break. As you have more than likely noticed the old menu bars are not there anymore. Before you break down and the tears start to roll, let's see what we can do.

Microsoft gave us some options for the keyboard user. Although the menus are not there, you can use the keyboard to access the different parts of the Ribbon. Not only can you access the Ribbon, but the old shortcuts (using the CTRL button) you have become accustomed to are still there and still working.

A complete list of the available shortcuts is on your computer in the Help section. If you want to see the complete list, click on the help button and type keyboard shortcuts in the search box. The Help button is the small question mark on the upper right side of the screen. Part of this list is reproduced below.

Press This	To Do This
CTRL+SHIFT+(	Unhides any hidden rows within the selection.
CTRL+SHIFT+)	Unhides any hidden columns within the selection.
CTRL+SHIFT+&	Applies the outline border to the selected cells.
CTRL+SHIFT_	Removes the outline border from the selected cells.
CTRL+SHIFT+~	Applies the General number format.
CTRL+SHIFT+$	Applies the Currency format with two decimal places (negative numbers in parentheses).
CTRL+SHIFT+%	Applies the Percentage format with no decimal places.

CTRL+SHIFT+^	Applies the Exponential number format with two decimal places.
CTRL+SHIFT+#	Applies the Date format with the day, month, and year.
CTRL+SHIFT+@	Applies the Time format with the hour and minute, and AM or PM.
CTRL+SHIFT+!	Applies the Number format with two decimal places, thousands separator, and minus sign (-) for negative values.
CTRL+SHIFT+*	Selects the current region around the active cell (the data area enclosed by blank rows and blank columns). In a PivotTable, it selects the entire PivotTable report.
CTRL+SHIFT+:	Enters the current time.
CTRL+SHIFT+"	Copies the value from the cell above the active cell into the cell or the Formula Bar.
CTRL+SHIFT+Plus (+)	Displays the **Insert** dialog box to insert blank cells.
CTRL+Minus (-)	Displays the **Delete** dialog box to delete the selected cells.
CTRL+;	Enters the current date.
CTRL+`	Alternates between displaying cell values and displaying formulas in the worksheet.

CTRL+'	Copies a formula from the cell above the active cell into the cell or the Formula Bar.
CTRL+1	Displays the **Format Cells** dialog box.
CTRL+2	Applies or removes bold formatting.
CTRL+3	Applies or removes italic formatting.
CTRL+4	Applies or removes underlining.
CTRL+5	Applies or removes strikethrough.
CTRL+6	Alternates between hiding objects, displaying objects, and displaying placeholders for objects.
CTRL+8	Displays or hides the outline symbols.
CTRL+9	Hides the selected rows.
CTRL+0	Hides the selected columns.

CTRL+A	Selects the entire worksheet. If the worksheet contains data, CTRL+A selects the current region. Pressing CTRL+A a second time selects the current region and its summary rows. Pressing CTRL+A a third time selects the entire worksheet. When the insertion point is to the right of a function name in a formula, displays the **Function Arguments** dialog box. CTRL+SHIFT+A inserts the argument names and parentheses when the insertion point is to the right of a function name in a formula.
CTRL+B	Applies or removes bold formatting.
CTRL+C	Copies the selected cells. CTRL+C followed by another CTRL+C displays the Clipboard.
CTRL+D	Uses the **Fill Down** command to copy the contents and format of the topmost cell of a selected range into the cells below.
CTRL+F	Displays the **Find and Replace** dialog box, with the **Find** tab selected. SHIFT+F5 also displays this tab, while SHIFT+F4 repeats the last **Find** action. CTRL+SHIFT+F opens the **Format Cells** dialog box with the **Font** tab selected.
CTRL+G	Displays the **Go To** dialog box. F5 also displays this dialog box.

CTRL+H	Displays the **Find and Replace** dialog box, with the **Replace** tab selected.
CTRL+I	Applies or removes italic formatting.
CTRL+K	Displays the **Insert Hyperlink** dialog box for new hyperlinks or the **Edit Hyperlink** dialog box for selected existing hyperlinks.
CTRL+N	Creates a new, blank workbook.
CTRL+O	Displays the **Open** dialog box to open or find a file. CTRL+SHIFT+O selects all cells that contain comments.
CTRL+P	Displays the **Print** dialog box. CTRL+SHIFT+P opens the **Format Cells** dialog box with the **Font** tab selected.
CTRL+R	Uses the **Fill Right** command to copy the contents and format of the leftmost cell of a selected range into the cells to the right.
CTRL+S	Saves the active file with its current file name, location, and file format.
CTRL+T	Displays the **Create Table** dialog box.
CTRL+U	Applies or removes underlining. CTRL+SHIFT+U switches between expanding and collapsing of the formula bar.

CTRL+V	Inserts the contents of the Clipboard at the insertion point and replaces any selection. Available only after you have cut or copied an object, text, or cell contents. CTRL+ALT+V displays the **Paste Special** dialog box. Available only after you have cut or copied an object, text, or cell contents on a worksheet or in another program.
CTRL+W	Closes the selected workbook window.
CTRL+X	Cuts the selected cells.
CTRL+Y	Repeats the last command or action, if possible.
CTRL+Z	Uses the **Undo** command to reverse the last command or to delete the last entry that you typed. CTRL+SHIFT+Z uses the **Undo** or **Redo** command to reverse or restore the last automatic correction when AutoCorrect Smart Tags are displayed.

Table 1-1

The following table lists the available Function keys.

F1	Displays the Microsoft Office Excel Help task pane. CTRL+F1 displays or hides the Ribbon, a component of the Microsoft Office Fluent user interface. ALT+F1 creates a chart of the data in the current range. ALT+SHIFT+F1 inserts a new worksheet.

F2	Edits the active cell and positions the insertion point at the end of the cell contents. It also moves the insertion point into the Formula Bar when editing in a cell is turned off. SHIFT+F2 adds or edits a cell comment. CTRL+F2 displays the Print Preview window.
F3	Displays the **Paste Name** dialog box. SHIFT+F3 displays the **Insert Function** dialog box.
F4	Repeats the last command or action, if possible. CTRL+F4 closes the selected workbook window.
F5	Displays the **Go To** dialog box. CTRL+F5 restores the window size of the selected workbook window.
F6	Switches between the worksheet, Ribbon, task pane, and Zoom controls. In a worksheet that has been split (**View** menu, **Manage This Window**, **Freeze Panes**, **Split Window** command), F6 includes the split panes when switching between panes and the Ribbon area. SHIFT+F6 switches between the worksheet, Zoom controls, task pane, and Ribbon. CTRL+F6 switches to the next workbook window when more than one workbook window is open.
F7	Displays the **Spelling** dialog box to check spelling in the active worksheet or selected range. CTRL+F7 performs the **Move** command on the workbook window when it is not maximized. Use the arrow keys to move the window, and when finished press ENTER, or ESC to cancel.

F8	Turns extend mode on or off. In extend mode, **Extended Selection** appears in the status line, and the arrow keys extend the selection.
	SHIFT+F8 enables you to add a nonadjacent cell or range to a selection of cells by using the arrow keys.
	CTRL+F8 performs the **Size** command (on the **Control** menu for the workbook window) when a workbook is not maximized.
	ALT+F8 displays the **Macro** dialog box to create, run, edit, or delete a macro
F9	Calculates all worksheets in all open workbooks.
	SHIFT+F9 calculates the active worksheet.
	CTRL+ALT+F9 calculates all worksheets in all open workbooks, regardless of whether they have changed since the last calculation.
	CTRL+ALT+SHIFT+F9 rechecks dependent formulas, and then calculates all cells in all open workbooks, including cells not marked as needing to be calculated.
	CTRL+F9 minimizes a workbook window to an icon.
F10	Turns key tips on or off.
	SHIFT+F10 displays the shortcut menu for a selected item.
	ALT+SHIFT+F10 displays the menu or message for a smart tag. If more than one smart tag is present, it switches to the next smart tag and displays its menu or message.
	CTRL+F10 maximizes or restores the selected workbook window.
F11	Creates a chart of the data in the current range.
	SHIFT+F11 inserts a new worksheet.
	ALT+F11 opens the Microsoft Visual Basic Editor, in which you can create a macro by using Visual Basic for Applications (VBA).

F12	Displays the **Save As** dialog box.

Table 1-2

The next table shows the rest of the available shortcut keys

ARROW KEYS	Move one cell up, down, left, or right in a worksheet.
	CTRL+ARROW KEY moves to the edge of the current data region (data region: A range of cells that contains data and that is bounded by empty cells or datasheet borders) in a worksheet.
	SHIFT+ARROW KEY extends the selection of cells by one cell.
	CTRL+SHIFT+ARROW KEY extends the selection of cells to the last nonblank cell in the same column or row as the active cell, or if the next cell is blank, extends the selection to the next nonblank cell.
	LEFT ARROW or RIGHT ARROW selects the tab to the left or right when the Ribbon is selected. When a submenu is open or selected, these arrow keys switch between the main menu and the submenu. When a Ribbon tab is selected, these keys navigate the tab buttons.
	DOWN ARROW or UP ARROW selects the next or previous command when a menu or submenu is open. When a Ribbon tab is selected, these keys navigate up or down the tab group.
	In a dialog box, arrow keys move between options in an open drop-down list, or between options in a group of options.
	DOWN ARROW or ALT+DOWN ARROW opens a selected drop-down list.
BACKSPACE	Deletes one character to the left in the Formula Bar.
	Also clears the content of the active cell.
	In cell editing mode, it deletes the character to the left of the insertion point.

27

DELETE	Removes the cell contents (data and formulas) from selected cells without affecting cell formats or comments. In cell editing mode, it deletes the character to the right of the insertion point.
END	Moves to the cell in the lower-right corner of the window when SCROLL LOCK is turned on.

Also selects the last command on the menu when a menu or submenu is visible.

CTRL+END moves to the last cell on a worksheet, in the lowest used row of the rightmost used column. If the cursor is in the formula bar, CTRL+END moves the cursor to the end of the text.

CTRL+SHIFT+END extends the selection of cells to the last used cell on the worksheet (lower-right corner). If the cursor is in the formula bar, CTRL+SHIFT+END selects all text in the formula bar from the cursor position to the end—this does not affect the height of the formula bar. |
| ENTER | Completes a cell entry from the cell or the Formula Bar, and selects the cell below (by default).

In a data form, it moves to the first field in the next record.

Opens a selected menu (press F10 to activate the menu bar) or performs the action for a selected command.

In a dialog box, it performs the action for the default command button in the dialog box (the button with the bold outline, often the **OK** button).

ALT+ENTER starts a new line in the same cell.

CTRL+ENTER fills the selected cell range with the current entry.

SHIFT+ENTER completes a cell entry and selects the cell above. |

ESC	Cancels an entry in the cell or Formula Bar.
	Closes an open menu or submenu, dialog box, or message window.
	It also closes full screen mode when this mode has been applied, and returns to normal screen mode to display the Ribbon and status bar again.
HOME	Moves to the beginning of a row in a worksheet.
	Moves to the cell in the upper-left corner of the window when SCROLL LOCK is turned on.
	Selects the first command on the menu when a menu or submenu is visible.
	CTRL+HOME moves to the beginning of a worksheet.
	CTRL+SHIFT+HOME extends the selection of cells to the beginning of the worksheet.
PAGE DOWN	Moves one screen down in a worksheet.
	ALT+PAGE DOWN moves one screen to the right in a worksheet.
	CTRL+PAGE DOWN moves to the next sheet in a workbook.
	CTRL+SHIFT+PAGE DOWN selects the current and next sheet in a workbook.
PAGE UP	Moves one screen up in a worksheet.
	ALT+PAGE UP moves one screen to the left in a worksheet.
	CTRL+PAGE UP moves to the previous sheet in a workbook.
	CTRL+SHIFT+PAGE UP selects the current and previous sheet in a workbook

SPACEBAR	In a dialog box, performs the action for the selected button, or selects or clears a check box.

CTRL+SPACEBAR selects an entire column in a worksheet.

SHIFT+SPACEBAR selects an entire row in a worksheet.

CTRL+SHIFT+SPACEBAR selects the entire worksheet.

If the worksheet contains data, CTRL+SHIFT+SPACEBAR selects the current region. Pressing CTRL+SHIFT+SPACEBAR a second time selects the current region and its summary rows. Pressing CTRL+SHIFT+SPACEBAR a third time selects the entire worksheet.

When an object is selected, CTRL+SHIFT+SPACEBAR selects all objects on a worksheet.

ALT+SPACEBAR displays the **Control** menu for the Microsoft Office Excel window. |
| TAB | Moves one cell to the right in a worksheet.

Moves between unlocked cells in a protected worksheet.

Moves to the next option or option group in a dialog box.

SHIFT+TAB moves to the previous cell in a worksheet or the previous option in a dialog box.

CTRL+TAB switches to the next tab in dialog box.

CTRL+SHIFT+TAB switches to the previous tab in a dialog box. |

Table 1-3

Are you ready for even more good news? Microsoft has included new shortcuts with the Ribbon. Why you might ask. It is because this change brings two major advantages. First there are shortcuts for every single button on the Ribbon and second because the many of the shortcuts require fewer keys.

The new shortcuts also have a new name: **Key Tips**

Using the keyboard press the Alt key

Pressing the Alt key will cause the **Key Tip Badges** to appear for all Ribbon tabs, the Quick Access Toolbar commands, and the Microsoft Office Button. After the Key Tip Badges appear, you can press the corresponding letter or number on the badge for the tab or the command you want to use. As an example, if you pressed Alt and then H you would bring the Home tab to the front. Figure 1-23 shows what the Ribbon looks like after pressing the Alt key.

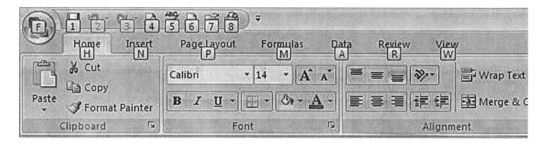

Figure 1-23

Note: You can still use the old Alt + shortcuts that accessed the menus and commands in the previous versions of Excel, but because the old menus are not available, you will have no screen reminders of what letters to press, so you will need to know the full shortcut to be able to use them.

But that is not all there is for the keyboard user. Microsoft has included shortcut menus. These are menus that you can access by right-clicking on a cell. The shortcut menu is shown in Figure 1-24.

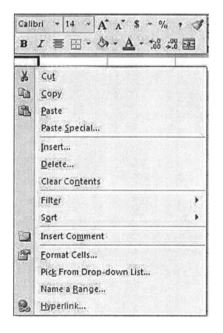

Figure 1-24

31

This shortcut menu shows all of the things that you can do to this cell. The bottom part looks a lot like the Edit from the previous version's menu bar. You will also notice the different formatting options that you can perform from the shortcut menu. See I told you this version is cool!

Lesson 1 – 7   Cell Names

When you are working with a spreadsheet you will have to reference cells in your formulas. Each cell in the spreadsheet has a unique name that identifies its location in the spreadsheet. The cell names are referenced by the column and the row where the cell is located. Using the cell name will make your formulas easier to understand and maintain.

The columns start from the left and move to the right. The first column is labeled A, the second is B, etc. The next column after column Z is column AA followed by AB. This continues on until column ZZ the next column will be column AAA and continues on, until you get to the maximum number of 16,384 columns. The rows start with the top row being number 1 and the second row being number 2 and this continues on until you get to the maximum number of 1,048,576 rows.

Table 1-4 shows the first few cell names.

A1	B1	C1	D1	E1	F1	G1	H1
A2	B2	C2	D2	E2	F2	G2	H2
A3	B3	C3	D3	E3	F3	G3	H3
A4	B4	C4	D4	E4	F4	G4	H4
A5	B5	C5	D5	E5	F5	G5	H5
A6	B6	C6	D6	E6	F6	G6	H6
A7	B7	C7	D7	E7	F7	G7	H7
A8	B8	C8	D8	E8	F8	8G	H8

Table 1-4

Using cell names may not appeal to you right now, but later as we continue, things will make a lot more sense. You will find out that you do not have any problem using cell references in your formulas. You may also decide that it would be almost impossible for Excel to work properly if it were not for the cell names.

Lesson 1 – 8   Using Labels and Text

If you were using a spreadsheet to keep tract of income and expenses for the year, you wouldn't randomly enter numbers into the spreadsheet. You may want each column to be for a specific type of income or expense. You may also want each row to represent a specific time frame, such as a week or a month. To keep everything from getting mixed up and to keep you from forgetting which column and row represented which item, you would want to use labels and text in your workbook.

There are two types of information that you can enter into a cell: Labels and Values. Values are discussed in the next lesson. In this lesson we will be inserting labels into our worksheet.

A label is any type of text or information that is not used in any calculations.

Open a new blank workbook

You can use the Quick Access Toolbar to open a new workbook by clicking on the "New" button. You can also select "New" and then Blank Workbook from the Office button.

Select cell A1 and type Month in it and then press Enter

You can select the cell by clicking it with the mouse. The first cell (A1) should be selected by default when you open the new worksheet. You can tell that it is the selected (active) cell by the black border around it.

When you press the enter key on the keyboard, the active cell will change to cell A2 or B1 depending on the setting you have in the Excel Options on the advanced tab.

In cell B1 type Rent and press the Tab key on the keyboard

Pressing the Tab key should make the next cell to the right the active cell.

By default, pressing the Enter key should cause the active cell to move one cell down and pressing the Tab key should cause the active cell to move one cell to the right.

Fill in the other cells as shown in Figure 1-25. You can select a cell by clicking on it with the mouse, or by using the Tab and Enter keys.

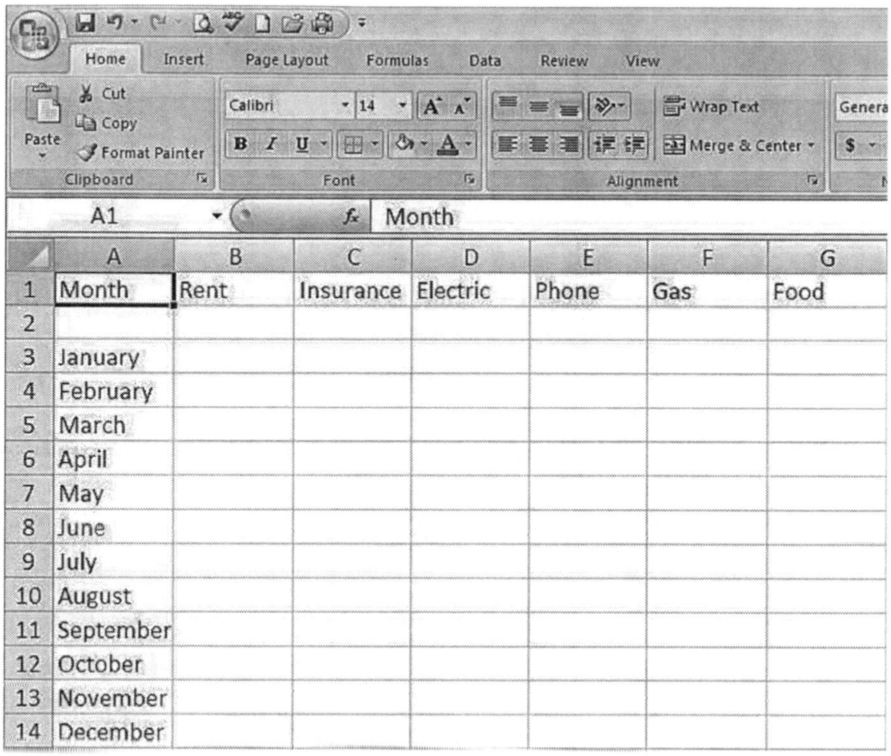

Figure 1-25

Each column (A through G) will represent a different type of category. Column A will have all of the months listed in it. The other columns will have the various expenses in them. Everything entered into column A is a label and everything entered into row 1 is also a label. In the next lesson we will work with some values.

Lesson 1 – 9   Entering Values

Now that we have some labels and we can keep everything straight, we will need to enter some data into our workbook. This data that we are going to enter is called a value.

A Value is any type of informational data: numbers, percentages, fractions, currencies, dates or times, usually used in formulas or calculations.

We will use our Family Budget that we started in the last lesson to enter our values.

In cell B3 enter 575 and press the Enter key.

You can select the cell to make it the active cell by clicking on it with the mouse. To enter the 575, you can use the numbers at the top of the keyboard or you can use the numbers on the keypad on the right side of the keyboard, if the numbers lock is turned on.

Pressing the Enter key will make the next cell down the active cell. This cell (B4) is now ready for you to enter a value into it.

Every month our rent should stay the same, so continue to fill the rest of the year in with 575 under the rent column. The result should look like Figure 1-26.

	A	B	C	D	E	F	G
1	Month	Rent	Insurance	Electric	Phone	Gas	Food
2							
3	January	575					
4	February	575					
5	March	575					
6	April	575					
7	May	575					
8	June	575					
9	July	575					
10	August	575					
11	September	575					
12	October	575					
13	November	575					
14	December	575					

Figure 1-26

36

Select cell C3 and enter 195 and then press Enter

This will show that our January insurance payment is 195. Our insurance payment should remain the same, so the rest of the year can be entered.

Fill in the remaining months so they will also show an insurance payment of 195

Now that you are getting into this, fill out the rest of the workbook until it looks like Figure 1-27.

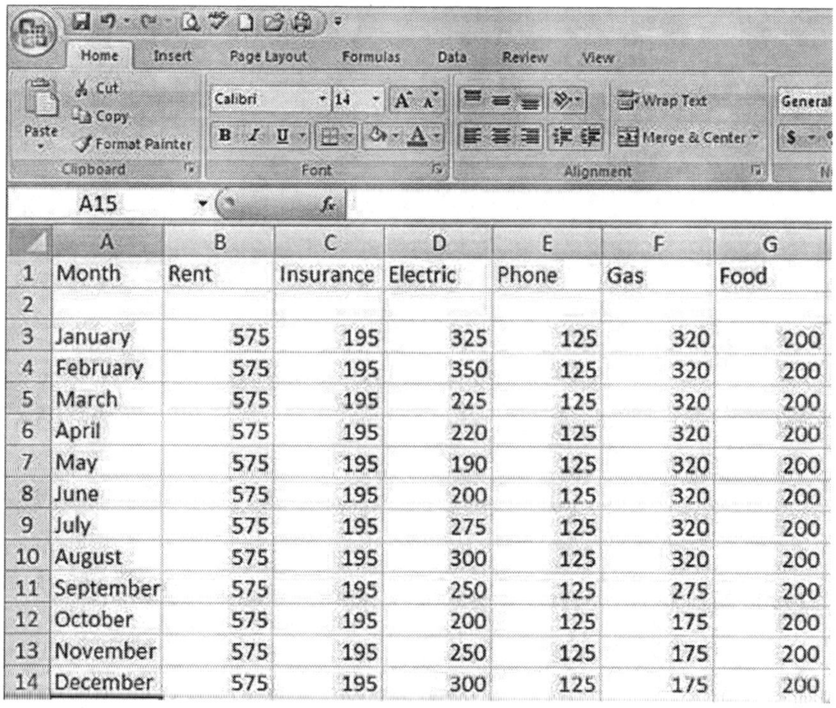

	A	B	C	D	E	F	G
1	Month	Rent	Insurance	Electric	Phone	Gas	Food
2							
3	January	575	195	325	125	320	200
4	February	575	195	350	125	320	200
5	March	575	195	225	125	320	200
6	April	575	195	220	125	320	200
7	May	575	195	190	125	320	200
8	June	575	195	200	125	320	200
9	July	575	195	275	125	320	200
10	August	575	195	300	125	320	200
11	September	575	195	250	125	275	200
12	October	575	195	200	125	175	200
13	November	575	195	250	125	175	200
14	December	575	195	300	125	175	200

Figure 1-27

Up to this point you have entered labels into our worksheet and you have entered some values into it also. There is so much more to tell you, so I guess we should go on to the next lesson.

Lesson 1 – 10   Saving the Workbook

Now that you have gone through all of the trouble of entering the information, we need to save it so nothing will happen to our data. I don't know about you, but I have gone through situations like this only to have the power go off and everything is gone. Before that happens, lets learn how to save the workbook.

At this point I am going to make an assumption. The assumption is that the workbook from the previous two lessons is still open. If it is not open, open it now.

Click the Office Button

When you click the Office button a drop down menu will appear as shown in Figure 1-28.

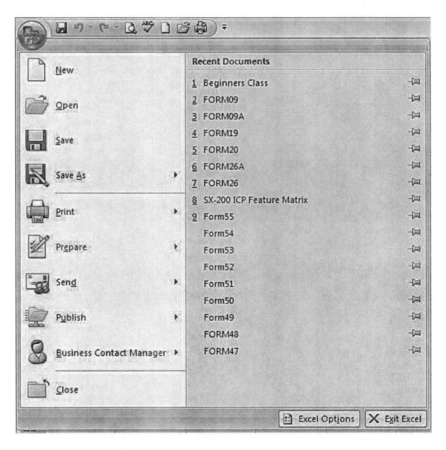

Figure 1-28

Note: the recent documents side of yours will not look like the one in figure 1-28; since these are my recent documents not yours.

You will notice that there are two save choices. Now would be a good time to explain the difference between the two of them. If this is the first time you have saved the current spreadsheet, The Save choice will bring the Save As dialog box to come to the screen. If this is not the first time you have saved the workbook the choice will have different results.

Save: This will replace the existing workbook with the newer version that is displayed on your screen. All changes made will be saved and the original workbook, before you made any changes will be gone.

Save As: This will allow you to save the currently displayed workbook with a different name. This will allow you to keep the original workbook just as it was before any changes were made to it and the new version will also be saved only under a different name.

Move the mouse to the Save As choice

Another menu will slide out to the side.

The first thing you have to do is make an important decision: what format do I want to save this workbook in. Figure 1-29 shows the choices you have to choose from.

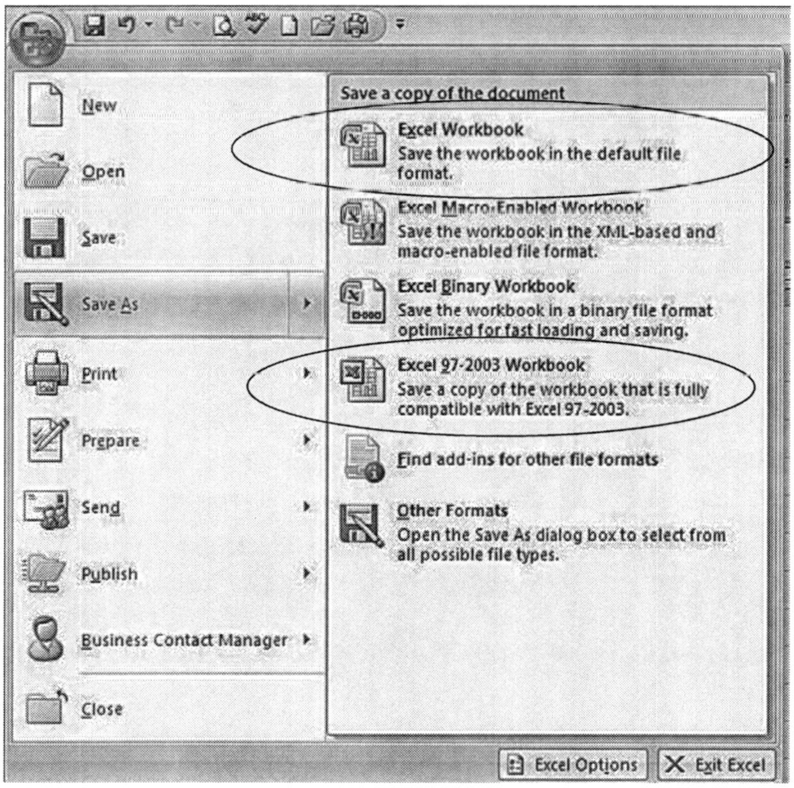

Figure 1-29

39

The two most obvious choices are: Save in the default format, which is Excel 2007 and Save in the older format of Excel 97 – 2003. If you were going to share this with someone who does not have Excel 2007 on their computer but has an older version of Excel, you would choose the Excel 97-2003 workbook choice.

Choose the top choice Excel Workbook

This will bring the Save As dialog box to the screen. Again, if this is the first time you have saved this workbook the Save As dialog box will also come to the screen. The reason for this is because the first time you save a workbook you have to give it a name. If it already has a name and you select the Save choice it will save the new version. I had you choose the Save As choice just to make sure you know how to use the Save As dialog box.

The dialog box is shown in Figure 1-30.

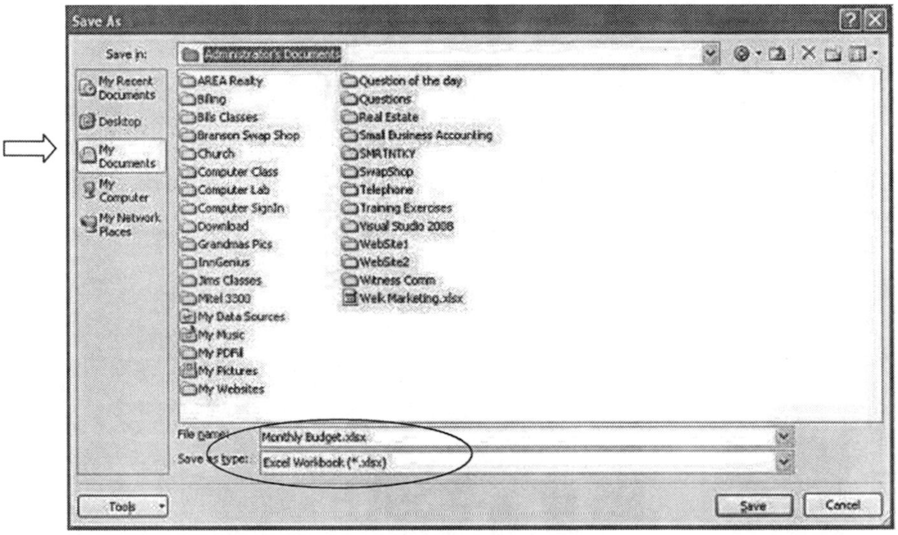

Figure 1-30

Your screen will look a little different than the one shown. The picture shown is of the document and folders on my computer, yours will show the documents and folders on your computer. For the time being, we want to save the workbook in the My Documents folder. When we get to chapter 5 we will move the file to a different folder.

Make sure My Documents is selected on the right side (if it is not click on it with the mouse).

Now all we have to do is give the workbook a unique name. It is essential that every file has a unique name. This will allow us to keep our computer organized, which we will discuss in chapter 5, and also allow us to find the file when we need it. The name should reflect something about the workbook, such as why it was created. This workbook is being used for our monthly budget, so a name like Monthly Budget is more practical than a name like Workbook 17.

In the text box next to file name type Monthly Budget and click the Save button

The dialog box will disappear from the screen and the workbook will be saved as an Excel 2007 file called Monthly Budget. You will be able to access this file any time you want by locating it under My Documents and double clicking it with the mouse. We will discuss opening an existing workbook in Lesson 1-12.

That is all there is to saving a file. Remember use Save As if you wish to keep the original workbook as it was before any changes were made and use Save if you wish to replace the original file with the revised version.

Note: From now on, when you are asked to save a workbook, make sure you save it in the <u>MY Documents</u> folder until you are told something different.

Lesson 1 – 11   Closing the Workbook

Now that we have saved our work we can safely close the file. This will be a very short lesson, as there is not very much to closing a workbook (file).

Click on the Office Button

You have seen this drop down menu before and it is shown again in Figure 1-31.

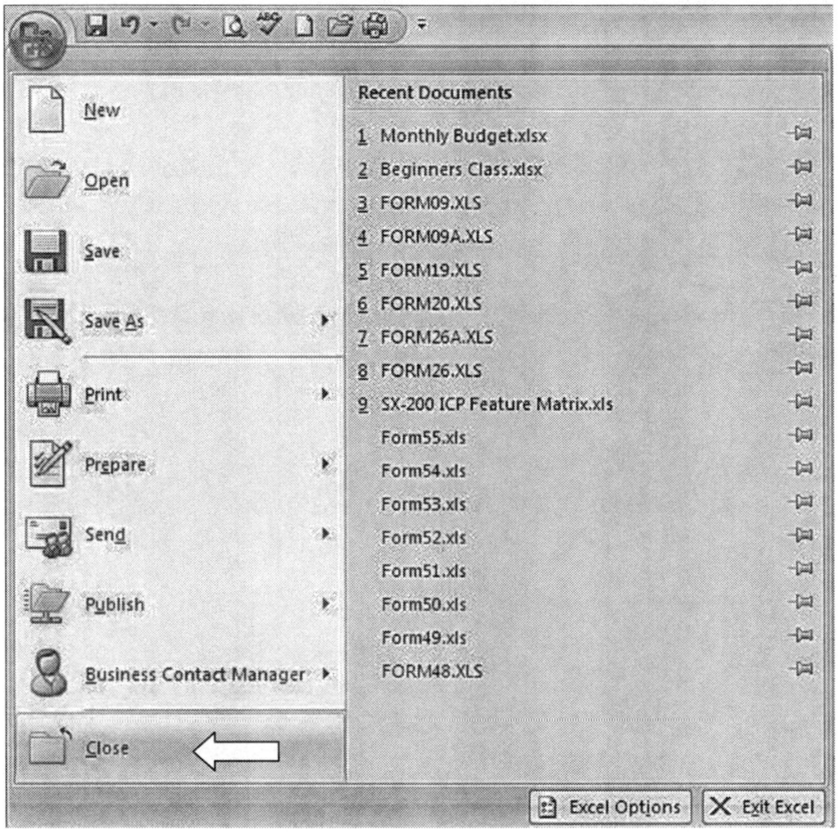

Figure 1-31

Click on the Close choice at the bottom

The workbook will close and you will end up with the main part of the screen being blank. In the next lesson we will see how to open the workbook.

Lesson 1 – 12   Opening a Workbook

If necessary open Excel

Remember that Excel is located in the Microsoft Office folder under all programs which is under the Start button.

There are three basic ways to open an existing Excel workbook. We will go through each method in this lesson. We will look at what is probably the most common way first.

Click the Office Button

You have seen this many times now and it is shown again in Figure 1-32.

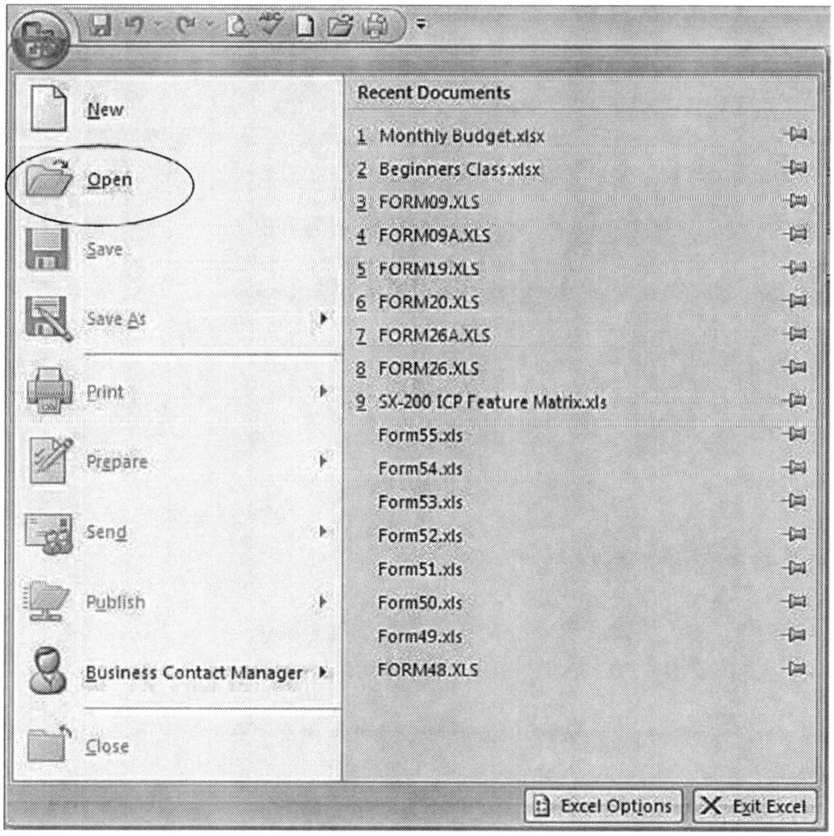

Figure 1-32

Click the Open button (the one with the circle around it)

This will bring the Open dialog box to the screen as shown in Figure 1-33.

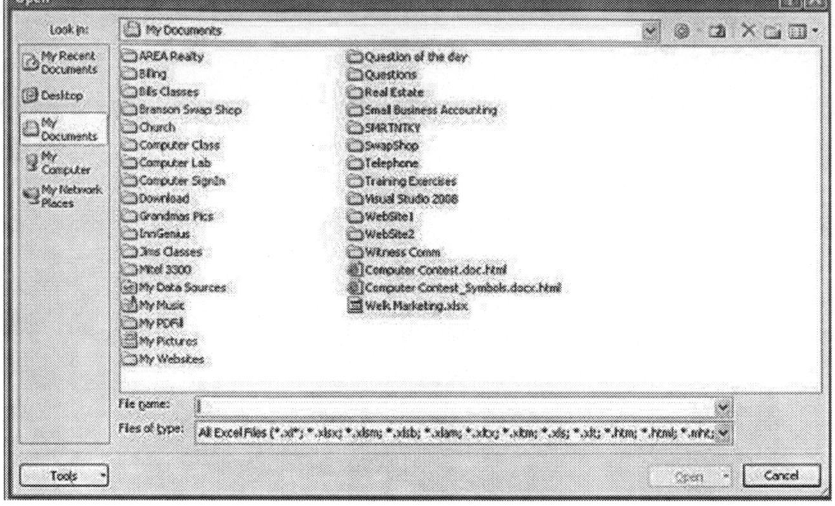

Figure 1-33

The normal starting place to look for a file is under My Documents, so that is the default place that should come up when you first see the Open dialog box. Since this is the place where we saved the file, you should see a file named Monthly Budget.xlsx. You will notice that the Open button is faded out and you cannot click on it at this point. Right now the button is disabled. As soon as you click on a file name the Open button will change and be enabled and be available to be clicked.

Click on Monthly Budget and then click the Open button

As soon as you click the Open button, Excel will start opening the workbook. It may take a second or two depending on the speed of your computer for the screen to actually change. But in a few moments the workbook will whisk onto your computer screen and you will be able to work with your spreadsheet.

Close the workbook as we did in lesson 1-11.

The second method for opening a workbook may be a little easier, but I am not sure that as many people use it as the first method we used.

Click on the Open button on the Quick Access Toolbar

The Quick Access Toolbar is show in Figure 1-34 and the open button is circled.

44

Figure 1-34

This will cause the Open Dialog box to immediately jump to the screen and you can continue just as we did on the previous page.

Click the Cancel button so nothing will open at this time

The third method for opening a file is about as easy as it gets if you have recently had the file open.

Click the Office Button to open the drop down menu as you did at the beginning of this lesson

You have seen this before. This time all we need to do is click on the name of the file on the right that we want to open. See Figure 1-35 to help explain this.

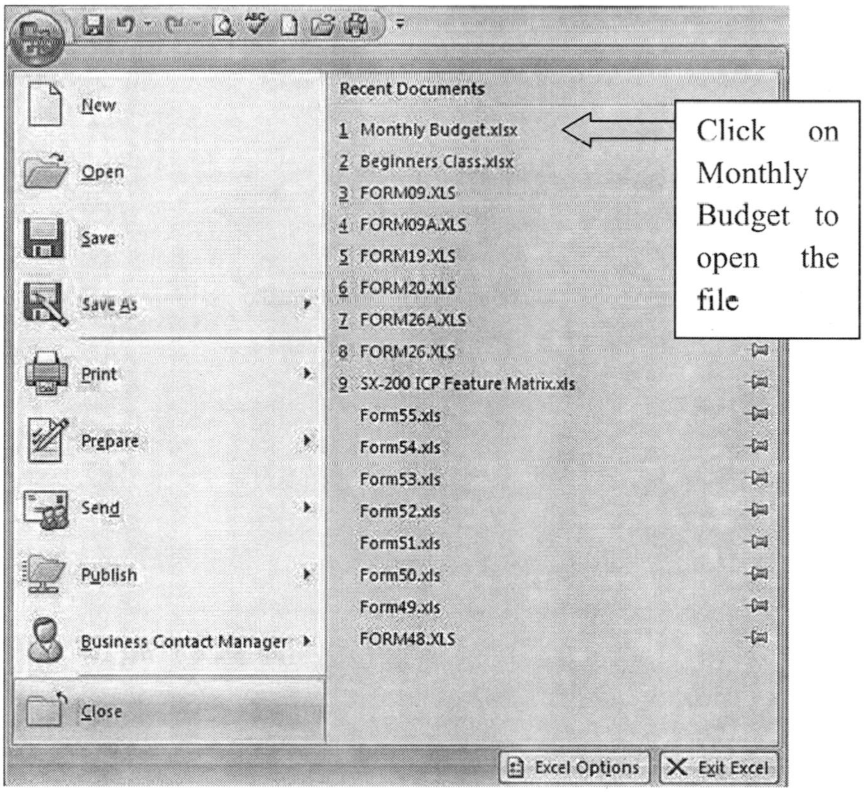

	Recent Documents	
New		
	1 Monthly Budget.xlsx	← Click on
	2 Beginners Class.xlsx	Monthly
Open	3 FORM09.XLS	Budget to
	4 FORM09A.XLS	open the
Save	5 FORM19.XLS	file
	6 FORM20.XLS	
Save As	7 FORM26A.XLS	
	8 FORM26.XLS	
Print	9 SX-200 ICP Feature Matrix.xls	
	Form55.xls	
Prepare	Form54.xls	
	Form53.xls	
Send	Form52.xls	
	Form51.xls	
Publish	Form50.xls	
	Form49.xls	
Business Contact Manager	FORM48.XLS	
Close		

Excel Options   X Exit Excel

Figure 1-35

45

Lesson 1 – 13   Printing a Workbook

Let's recap: We can enter labels, enter values, save the file, close the file, and open a file. Now let's see about printing a workbook. The workbook may be part of a report you are submitting or it may just be your personal budget. Either way you will probably want a printed copy. Printing the workbook is a simple process, provided that you have a printer set up on your computer.

If it is not open, open the workbook named Monthly Budget

The workbook should be displayed on the screen and look like it did when we last worked on it.

Click the Office Button and move the mouse to the Print choice, but <u>do not</u> click the mouse

There are three choices you can make from this menu: Print, Quick Print, and Print Preview. Before we actually print the workbook, it might be a good idea to see how it will look when we print it. We may find out that it will not all fit on one page and we may have to do some adjusting to our sheet. The available print choices can be seen In Figure 1-36.

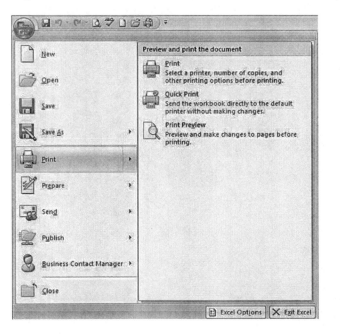

Figure 1-36

Click on the Print Preview choice

This will display how the sheet will look when it is printed. There are also a few other options available from this screen (see Figure 1-37).

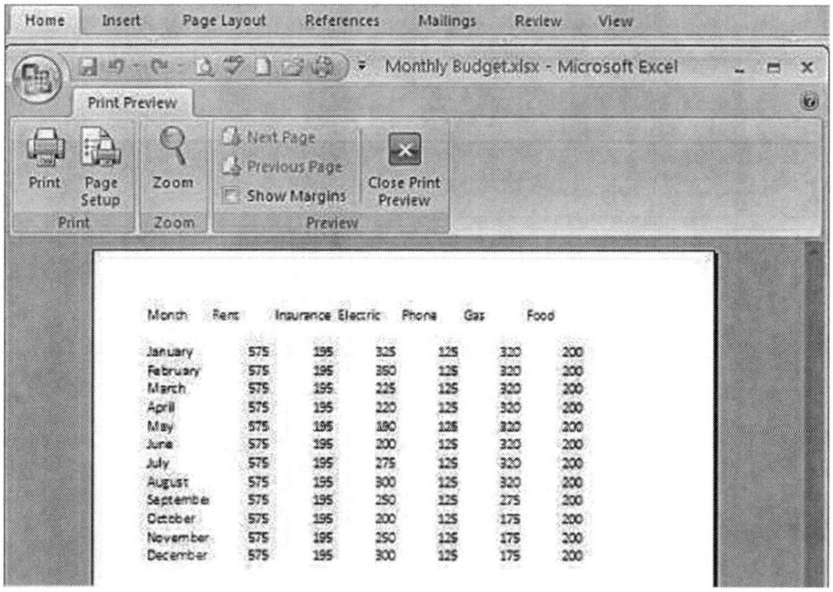

Figure 1-37

From this screen you can: Close the print preview, view the other pages (if there are any other pages), Zoom in or out on the document, make page settings, or print the workbook.

In the Preview group click the checkbox next to Show Margins

This will show you where the margins and tab stops are positioned. When you are finished viewing these, click the checkbox a second time to unselect it.

Click the Zoom button once to make the sheet larger

This will only make it larger on the computer screen. It will not actually make the spreadsheet larger. Click the button a second time to return the sheet back to its original size.

Click on the Page Setup button

This will bring the Page Setup dialog box to the screen. The first tab of the dialog box is the Page tab. If it is not on the front and showing, click on it. If it is at the front, the dialog box should look like Figure 1-38.

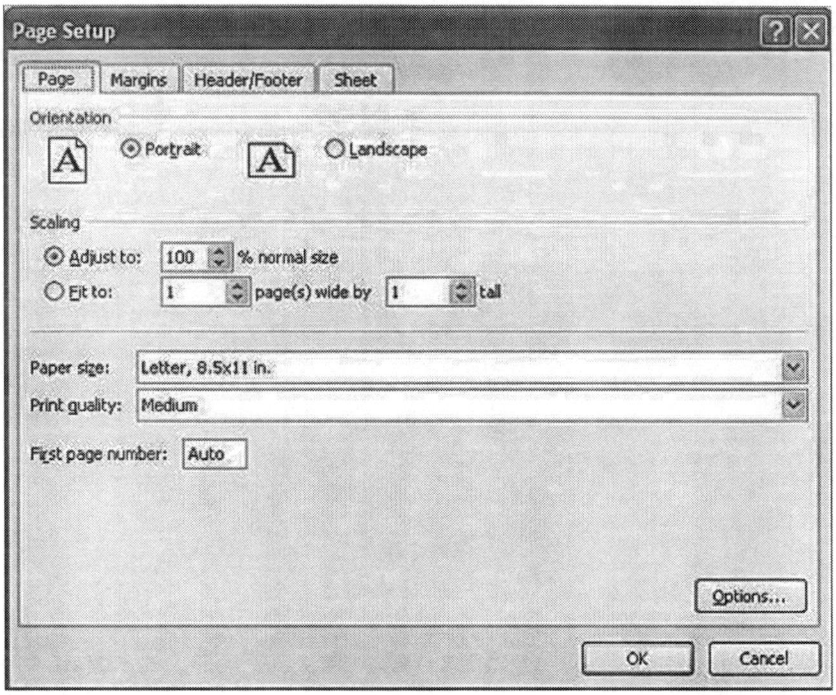

Figure 1-38

Some of the things you can change from this screen include: The page orientation (Landscape and Portrait), the scaling, the paper size, and print quality. If you are not sure of the difference between the two orientation types, I will explain. Portrait has the long side of the paper going up and down while Landscape has the long side of the paper going across the top and bottom.

The Options button will allow you to change most of the default printing settings. You should not have to change these, so I recommend that you don't.

Click on the Margins Tab

This will allow you to change the settings for the margins of the paper. You can also center the spreadsheet on the paper. Figure 1-39 shows the Margins Tab

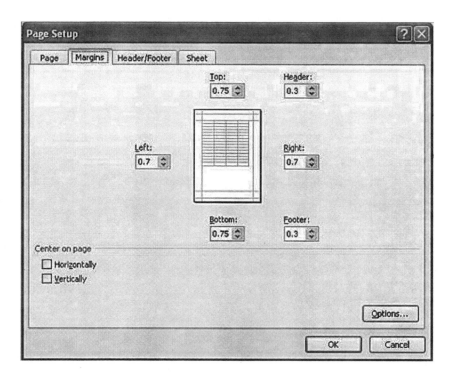

Figure 1-39

Click on the Header/Footer tab

Now you can decide what the text will be for the header and the footer. You can choose from the standard ones suggested by Excel, or you can choose to create your own custom header and footer by clicking on the Custom button. Figure 1-40 shows the Header/Footer tab.

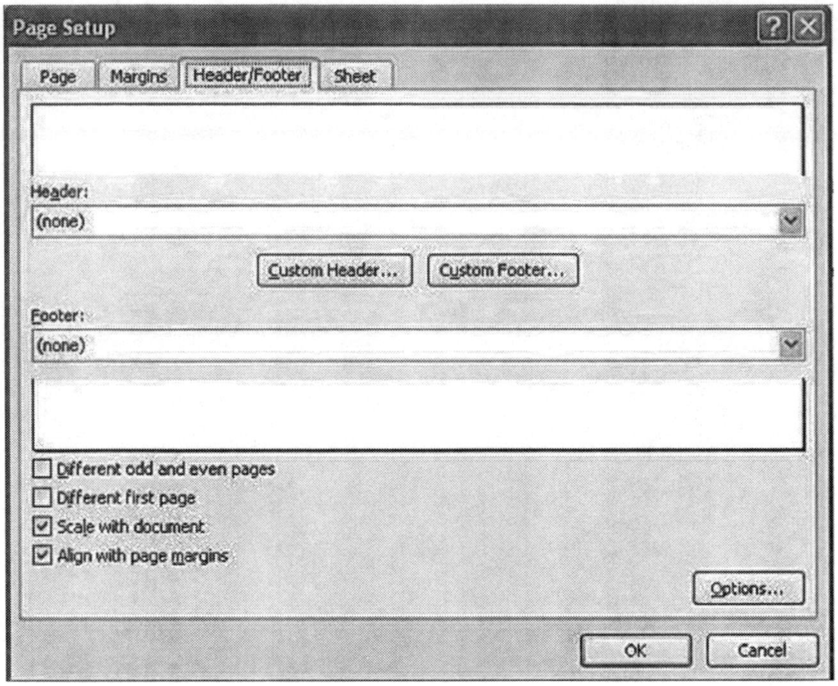

Figure 1-40

Click the Sheet Tab

There are more choices here. The only one you might want to change is the Gridlines checkbox. This will determine if there is going to be an outline around each cell. You will probably want this turned on.

In the Print section click on the checkbox next to Gridlines and then click the OK button

When you click the OK button the print preview will change to show the gridlines of each cell, and they will be printed as well. Figure 1-41 shows the Sheet tab and Figure 1-42 shows what the sheet will look like after you clicked the OK button.

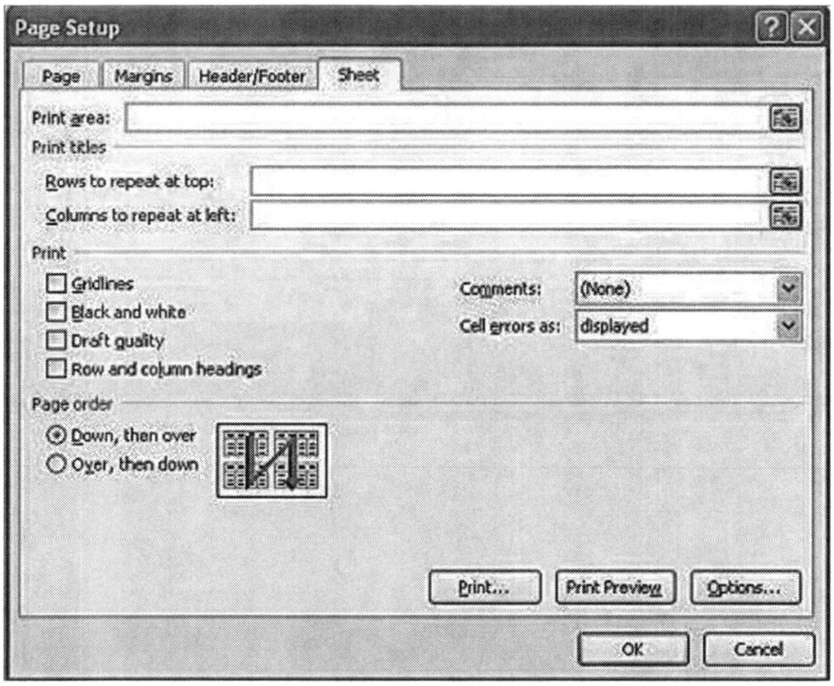

Figure 1-41

Month	Rent	Insurance	Electric	Propane	Phone	Gas	Food
January	575	195	325	125	125	320	200
February	575	195	350	130	125	320	200
March	575	195	225	125	125	320	200
April	575	195	220	110	125	320	200
May	575	195	190	65	125	320	200
June	575	195	200	35	125	320	200
July	575	195	275	40	125	320	200
August	575	195	300	35	125	320	200
Septembe	575	195	250	75	125	275	200
October	575	195	200	75	125	175	200
November	575	195	250	95	125	175	200
December	575	195	300	125	125	175	200

Figure 1-42

Click the Close Print Preview button

Now that we have seen what it is going to look like when we print it, let's see what the other two choices are when we moved the mouse over the print option.

Click on the Office button and move the mouse down to the word Print and then click on the top choice: Print

This will bring the Print Dialog box to the screen and is shown in Figure 1-43.

51

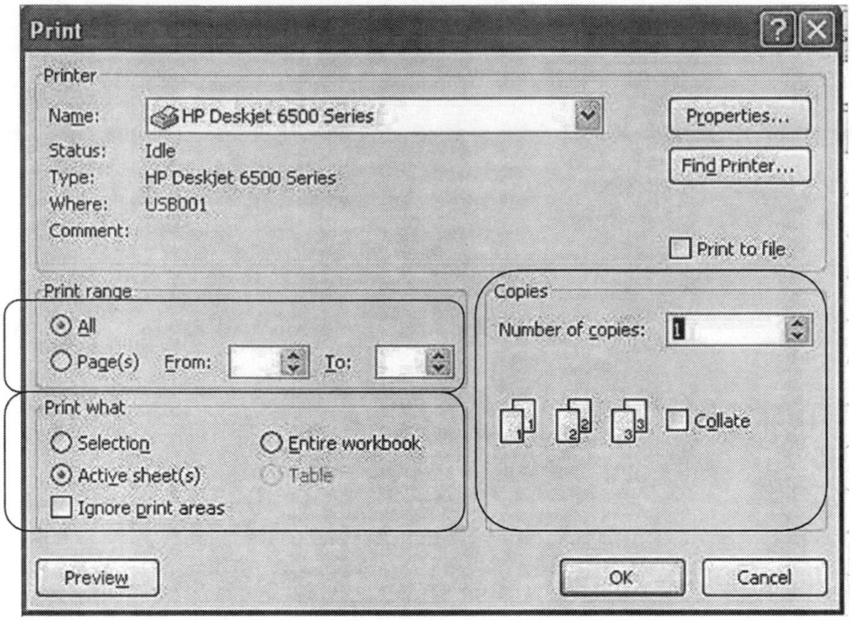

Figure 1-43

As you can see, we have a few choices that we can make. First, we can choose to print all of the pages in the top sheet or choose to only print the pages that we specify. The All Pages is chosen by default. If we do not want to print all of the pages, we can click the radio button next to the word Pages and then specify the "From" and "To" page numbers.

The above paragraph may have sounded strange and I will try to clarify it here. Let's assume that there are 75 rows and 10 columns in our worksheet (sheet 1). When we print this there will be three pages that print. We may not want all three pages to print, only the first page or perhaps the second page. This is how one sheet can have several pages on it and why we get to make this choice.

Directly below this is the "Print What" section. In this section we can choose to print only the selection that we have highlighted, print only from the active sheet, print the entire workbook (all sheets that have something on them), or ignore the print area. That last part makes no sense at all. One of the things that we have not covered in detail yet is the Ribbon. Using the Ribbon we can set what the print area actually is. Here we can choose to ignore the area we designated as the print area.

On the right side, we can choose how many copies we want to print. We can also choose to collate the pages (put them in the correct order as we print).

Once we have everything marked correctly, all we have to do is click on the OK button and this will send the information to the printer.

Click Cancel so nothing will print

We can always print this at any time and we are not finished with this workbook yet.

Chapter 1   Review

Starting Excel – Excel can be started by clicking the Start Button, All Programs, Microsoft Office, and then on Microsoft Excel 2007.

Understanding the Excel Screen – Be able to identify the following parts of the Excel Screen:

> Office Button
> Quick Access Toolbar
> The Ribbon
> The Formula Bar
> Cell Name Box
> The Active Cell
> Column Headings
> Row Headings

The Office Button – Know the main parts of the Office Button:
> New
> Open
> Save
> Save As
> Print
> Close

The Ribbon – Be familiar with the main parts of the Ribbon
> Tabs
> Groups
> Commands
> Dialog Boxes

Quick Access Toolbar – Provided for your convenience and ease of use

Keyboard Shortcuts – Still available and there for you to use
> Key Tips are now available using the ALT key plus the letter or number of the Key Tip Badge.
> Right-click on a cell or object will bring the Shortcut Menu to the screen.

The default cell name comes from the column and row the cell is located in.

Labels are information in cells that are not normally used in a calculation.

Values are information in cells that are normally used in calculations.

Save replaces the original file with the new version that is displayed on the screen, while Save As will allow you to keep the original and the new version by giving the new version a different name.

To open a workbook, click on the Office Button and choose Open from the menu.

Chapter 1   Quiz

1. Excel is only used for complicated mathematical calculations; simple math cannot be done with Excel. **True or False**

2. Name the Three main parts of the Ribbon.

3. How can you select cell A3
    A. Type C3 using 5the keyboard
    B. Press ALT and then type C3 using the keyboard
    C. Click on cell C3 with the mouse
    D. Press C and 3 at the same time

4. If a command is not seen on the Ribbon, it may be inside a Dialog Box. How do you open a Dialog Box?

5. Finding Key Tip Badges is a relatively complicated process. **True or False**

6. Default cell names have two parts; where do they come from?

7. Labels contain information that is used in calculations. **True or False**

8. Workbooks can only be saved in the new 2007 format. **True or False**

9. Clicking the Zoom button when you are viewing the Print Preview screen will cause the print to get larger every time you click the button. **True or False**

10. The Page Setup from the Print Preview screen will not allow you to change the page orientation. **True or False**

Chapter two   The Ribbon - A Closer Look

In chapter one, we had an overview of the Ribbon. In this chapter we will take a closer look at the different parts of the Ribbon.

In this chapter we will look at the different tabs of the Ribbon, and the groups on the Tab.

A word of concern: Depending on the size of your monitor, you may not see everything that is shown in the figures. Some of the groups may be condensed and may not show all of the available options at all times. The options are still available, but you may have to click one of the drop-down arrows to see them.

This chapter will not have much user interface and consists of mostly reading (sorry about that), but it is necessary to have an understanding of the Ribbon.

Open the workbook named Monthly Budget, if it is not open

Lesson 2 – 1   The Home Tab

The Home Tab of the Ribbon is shown in Figure 2-1.

Figure 2-1

Now let's take a closer look at each of the Groups on the Tab.

The first Group (on the far left) is the Clipboard group and is shown in Figure 2-2.

Figure 2-2

The figure on the left shows the view if the monitor screen is not wide enough to show all of the words. This group deals with the different things you can do with selected text, values, and cells. These things will be covered in detail in Chapters three and six.

The next Group is the Font Group and is shown in Figure 2-3.

Figure 2-3

As you would expect, this is where you would perform all of the formatting for the text. This will be discussed in Chapter six.

The next Group is the Alignment Group and is shown in Figure 2-4.

Figure 2-4

This is where you determine how the text is to be aligned inside the cell. You can also merge cells together form here.

The next Group is the Number Group and is shown in Figure 2-5.

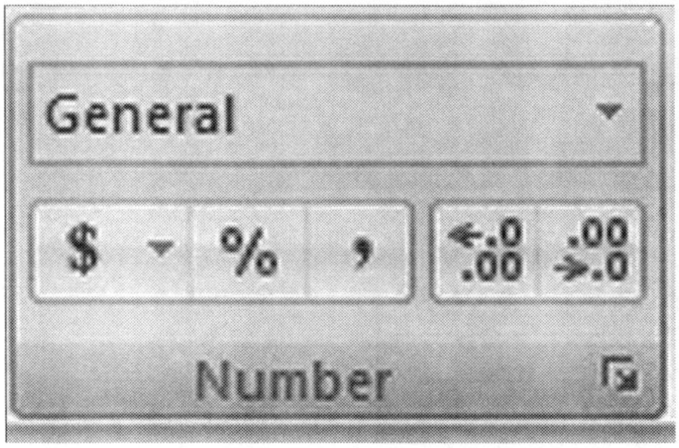

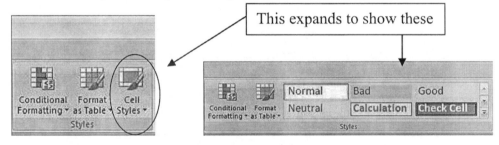

Figure 2-5

This group will let you decide how numbers are displayed in the cells, and the number of decimal points in a number.

The next Group is the Styles Group and is shown in Figure 2-6.

This expands to show these

Figure 2-6

From this group you can apply a style to the workbook. Styles will be discussed in Chapter seven.

The next Group is the Cells Group and is shown in Figure 2-7. Two different views of the same group are shown.

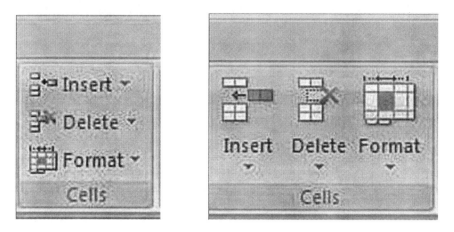

Figure 2-7

From within this group we can insert and delete cells as well as format the cells. There will be more on this in Chapter three.

The last group on the tab is the Editing Group and is shown in Figure 2-8. This also shows two views of the same group.

Figure 2-8

Using this group you can use the AutoSum, which will be covered in Chapter four. I bet you thought that this would be on the Formula Tab, since it deals with mathematical calculations. Well, surprise, it is on both tabs. Remember, I said the Home tab contained the things that you use the most. This is one of the buttons that you will use a lot so it is included on the Home Tab.

You can also use the fill feature which lets you continue a pattern down the column. This will also be covered in Chapter four. The clear button will remove anything in the cell or remove selected formatting. This is covered in Chapter three.

The Sort and Filtering will let you arrange the data so it will be easier to analyze. This is also covered in Chapter three.

The last button is the Find and Select button. This will allow you to find specific text or type of information. This will be covered in Chapter three.

Lesson 2 – 2   The Insert Tab

Click on the Insert Tab to view its contents

As the name implies, this tab contains objects that can be inserted into your workbook.

The first group on the tab is the Tables Group and is shown in Figure 2-9.

Figure 2-9

A PivotTable allows you to take large amounts of data analyze it and then summarize it in a report. A table is similar, but you don't get the report feature. Tables will be covered in Chapter nine.

The next tab is the Illustrations Tab and is shown in Figure 2-10.

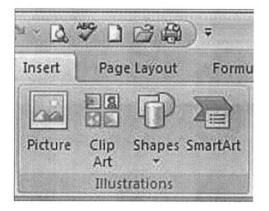

Figure 2-10

This is where you insert pictures and shapes into your workbook. This is covered in Chapter ten.

The next group is the Charts Group and is shown in Figure 2-11.

Figure 2-11

This is where you will insert chats into your workbook. This is covered in Chapter eight.

The next group is the Links Group and is shown in Figure 2-12.

Figure 2-12

This group will allow you to create a link between a cell and another workbook, web page, program, or picture. Hyperlinks will be covered in Chapter eleven.

The next group is the Text Group and is shown in Figure 2-13.

Figure 2-13

This is where you can insert a text box, symbols, WordArt, Headers and Footers, and other objects. This will be covered in Chapter ten.

Lesson 2-3   The Page Layout Tab

Click on the Page Layout Tab to view its contents

The first group on the Page Layout Tab is the Themes Group and is shown in Figure 2-14.

Figure 2-14

The themes group is where you can change the entire look of the workbook including the font and the colors. This will be briefly mentioned in Chapter seven.

The next group is the Page Setup Group and is shown in Figure 2-15.

Figure 2-15

From here you can change the margins, orientation, paper size, etc. and will be covered in Chapter thirteen.

The next group is the Scale to Fit Group and is shown in Figure 2-16.

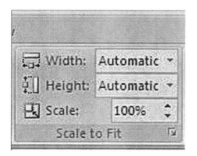

Figure 2-16

This is where you can adjust the height, width, and scale of the printed output. This will be discussed in Chapter thirteen.

The next group is the Sheet Options group and is shown in Figure 2-17.

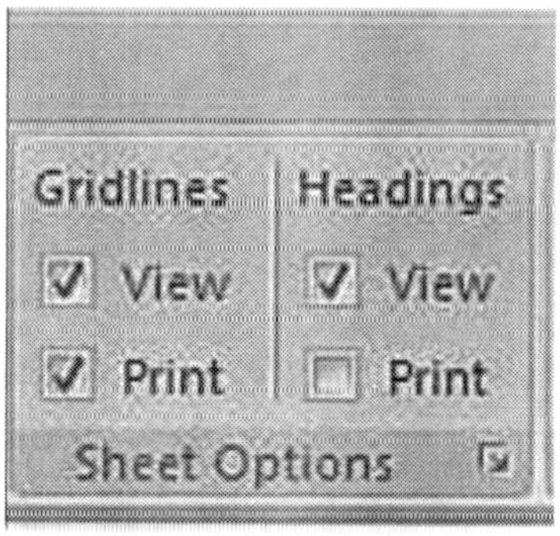

Figure 2-17

This is where you can decide if you are going to view and print the gridlines and headings. This will be covered in Chapter thirteen.

The last group in the Page Layout group is the Arrange Group (see Figure 2-18).

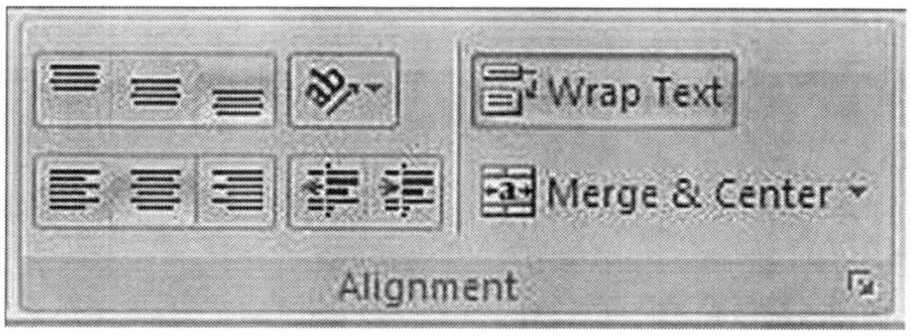

Figure 2-18

This deals with arranging object on the pages. Chapter thirteen will cover this.

Lesson 2 – 4   The Formulas Tab

Click on the Formulas Tab

As you can guess, this tab deals with Formula related commands.

The first group on the Formulas Tab is the Function Library Group and is shown in Figure 2-19.

Figure 2-19

As the name implies, this is a library of the available functions that are available for you to use. You can browse through the available functions by category. Some of these will be covered in Chapter four.

The next group is the Defined Names Group and is shown in Figure 2-20.

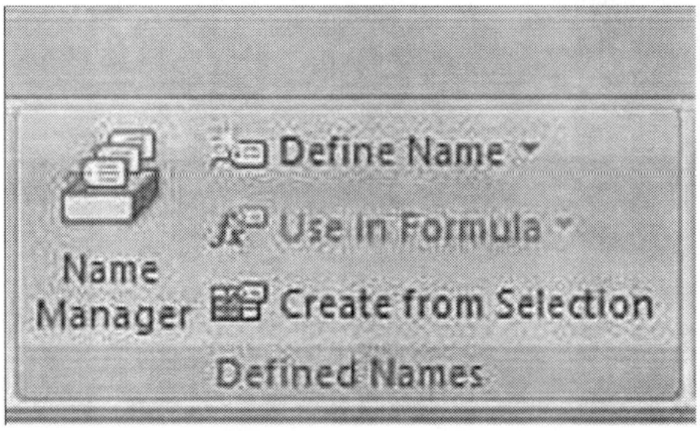

Figure 2-20

This group of commands will allow you to change the names of cells from the default names. This will be covered in Chapter four.

The next group is the Formula Auditing Group and is shown in Figure 2-21.

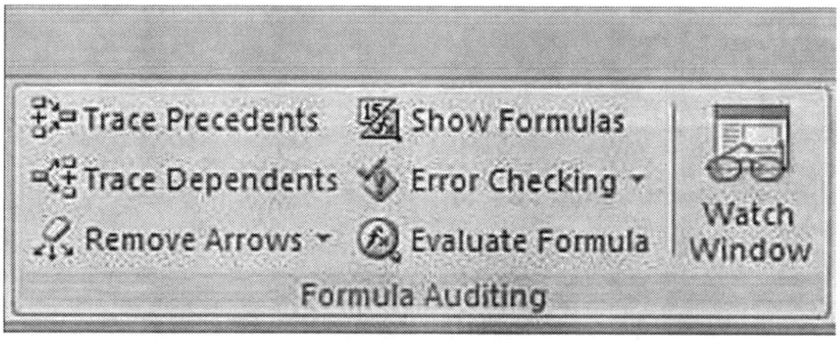

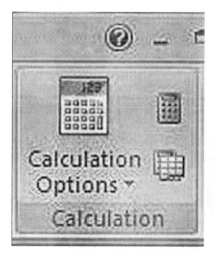

Figure 2-21

From here you can trace which cells affect the highlighted cell. You can also show which cells are affected by the highlighted cells. You can also show the formulas in the cell instead of the resulting value. We will discuss this more in Chapter four.

The last group is the Calculation Group and is shown in Figure 2-22.

Figure 2-22

From here you can choose if the calculations in the sheet are done automatically or manually. There will be more on this in Chapter four.

Lesson 2 – 5   The Data Tab

Click on the Data Tab

The first group on the Data Tab is the Get External Data Group and is shown in Figure 2-23.

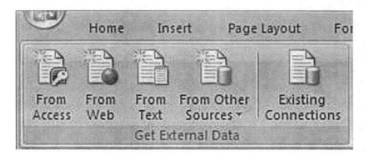

Figure 2-23

This is where you can choose where to get the data for the workbook. This can be from a Microsoft database, from the web, from a text file, etc. You could always enter the data into the workbook manually. Chapter nine will cover this.

The next group to the right is the Connections Group and is shown in Figure 2-24.

Figure 2-24

From here you can display all of the connections to an external data source. You can also refresh all of the information that is coming into the workbook from an external source. This will also be covered in Chapter nine.

The next group is the Sort and Filter Group and is shown in Figure 2-25.

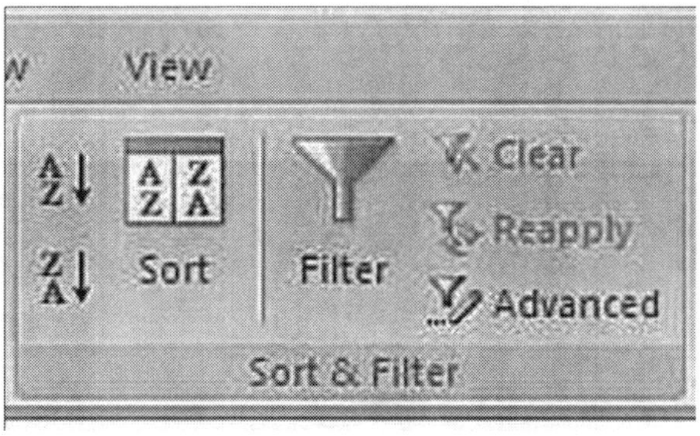

Figure 2-25

From here you can decide how your data should be sorted and if any filters are to be added to the sort. This will be covered in Chapter three.

The next group is the Data Tools Group and is shown in Figure 2-26.

Figure 2-26

From here you can separate the contents of one cell into separate columns. An example of this would be if one cell had both first and last names in it, you could separate the two names into two columns. You could also remove any duplicate entries using the data tools. You can also prevent invalid data from being entered into a cell using Data Validation. Some of this will be covered in Chapter three.

The last group on the Data Tab is the Outline Group and is shown in Figure 2-27.

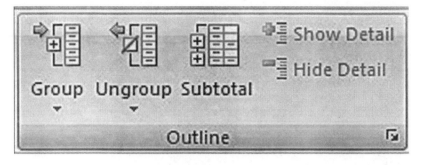

Figure 2-27

These commands will allow you to group, and ungroup, cells together and automatically inserting totals and subtotals for rows. This will be covered in Chapter four.

Lesson 2 – 6　　　The Review Tab

Click on the Review Tab

The first group on the Review Tab is the Proofing Group. This group is shown in Figure 2-28.

Figure 2-28

This group will give you tools you will need to proof your sheets. These tools include the spell checker, the Research Library, where you can search through dictionaries and encyclopedias. You also have access to the Thesaurus where you can search for similar words. The translate button will allow you to translate your text into different languages. Some of this will be covered in Chapter three.

The second group is the Comments Group and is shown in Figure 2-29.

Figure 2-29

Using these commands you can add comments to your cells. This will be covered in Chapter three.

The last group on the Review Tab is the Changes Group. This group is shown in Figure 2-30.

Figure 2-30

From here you can protect your sheet and workbook from having any changes made to it. You can also share the workbook from here and track any changes that have been made to the workbook. This will be covered in Chapter twelve.

Lesson 2 – 7   The View Tab

Click on the View Tab

The first group on the View Tab is the Workbook Views group. This is shown in Figure 2-31.

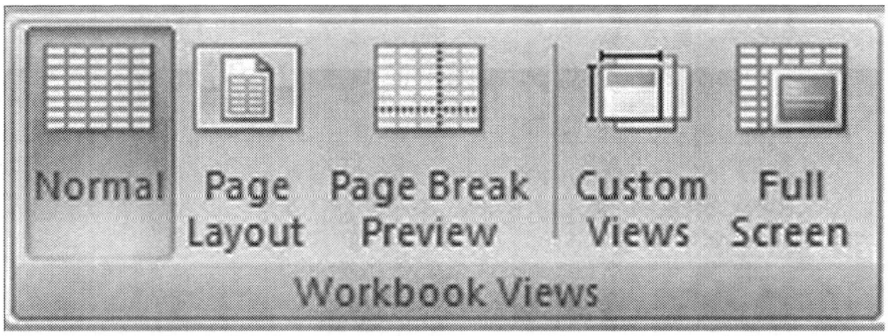

Figure 2-31

These commands let you change how the workbook is displayed on the screen. These views will be discussed in Chapter Fifteen.

The next group is the Show/ Hide Group and is shown in Figure 2-32.

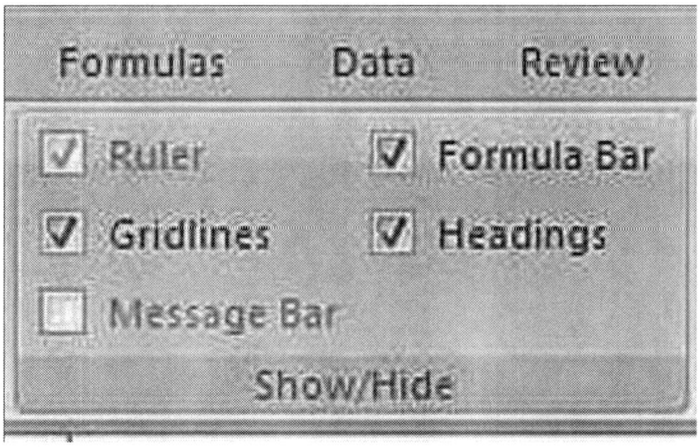

Figure 2-32

Using these commands you can decide what will be seen and printed in your workbook. This will be covered in Chapter Fifteen.

The next group is the Zoom Group and is shown in Figure 2-33.

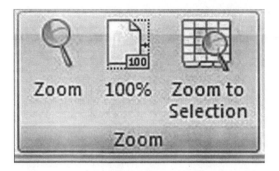

Figure 2-33

This group, as implied by the name, allows you to zoom in and out on the workbook. We will discuss this in Chapter Fifteen.

The next group is the Window Group and is shown in Figure 2-34.

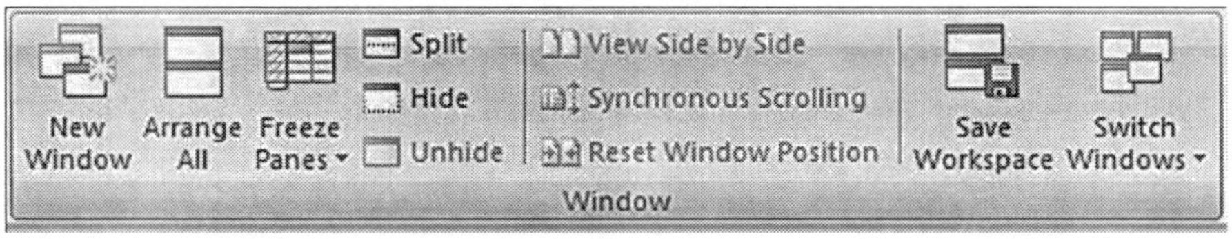

Figure 2-34

Using these commands we can change how the windows are displayed on the screen and also have some parts of the screen stay frozen while other parts move. We will be talking about this in Chapter Fifteen.

The last group on the View Tab is the Macro Group and is shown in Figure 2-35.

Figure 2-35

Macros are used when you find yourself typing the same thing over and over again. You can record keystrokes and insert them when needed. We will cover Macros in Chapter fourteen.

Chapter 2   Review

The Home Tab contains the following Groups:
    Clipboard
    Font
    Alignment
    Number
    Styles
    Cells
    Editing
The Insert Tab contains the following Groups:
    Tables
    Illustrations
    Charts
    Links
    Text
The Page Layout Tab contains the following Groups:
    Themes
    Page Setup
    Scale to Fit
    Sheet Options
    Alignment
The Formulas Tab contains the following Groups:
    Function Library
    Defined Names
    Formula Auditing
    Calculation
The Data Tab contains the following Groups:
    Get External Data
    Connections
    Sort & Filter
    Data Tools
    Outline
The Review Tab contains the following Groups
    Proofing
    Comments
    Changes

The View Tab contains the following Groups:
    Workbook Views
    Show / Hide
    Zoom
    Window
    Macros

Chapter 2   Quiz

1) What tab would you use to add a hyperlink to your workbook?
2) What tab would you use to add a macro to your workbook?
3) What tab would you use to change a cell name?
4) What tab would you use to sort the data in your workbook?
5) What tab would allow you to see which cells affect the selected cell?
6) What tab would you find the Sheet Options on?
7) What tab would you use to add a chart to your workbook?
8) What tab would you use to change the font formatting?
9) What tab has the copy and paste command on it?
10) What tab would you use to add comments to your workbook?

Chapter Three   Editing

After you have your labels and values entered into your workbook, the odds are that things will change. Now you will need to go back and make the changes into your worksheets. This chapter is dedicated to the various ways that you will use to edit existing data.

Lesson 3 – 1   Using Cut, Copy, and Paste

This lesson is dedicated to moving text and values around in your worksheet. We will be using Cut, Copy, and Paste to accomplish this. First, let me give you a brief description of these commands.

Cut: This will <u>remove</u> any data that is in the cell and move it to the clipboard. The clipboard is a temporary storage place and will temporarily hold the data until you can put it someplace else in either the worksheet or even into a completely different program.

Copy: The copy command will <u>make a copy</u> of the selected text (or value) and place it in the clipboard for you to use later.

Paste: The paste command will <u>copy from the clipboard</u> and put the information into the selected cell of the workbook.

Open the Monthly Budget workbook

We are going to use the Cut and Paste commands to move one of the columns in our worksheet.

Select the column food by clicking on the letter "G" in the column heading

The entire column will be highlighted in light blue, indicating that it is selected. While the column is selected we can use the Cut command to remove it from the worksheet and place it in the temporary storage container called the Clipboard. It should look like Figure 3-1.

	A	B	C	D	E	F	G	H
							Food	
1	Month	Rent	Insurance	Electric	Phone	Gas	Food	
2								
3	January	575	195	325	125	320	200	
4	February	575	195	350	125	320	200	
5	March	575	195	225	125	320	200	
6	April	575	195	220	125	320	200	
7	May	575	195	190	125	320	200	
8	June	575	195	200	125	320	200	
9	July	575	195	275	125	320	200	
10	August	575	195	300	125	320	200	
11	September	575	195	250	125	275	200	
12	October	575	195	200	125	175	200	
13	November	575	195	250	125	175	200	
14	December	575	195	300	125	175	200	
15								
16								

Figure 3-1

Click on the Home Tab of the Ribbon to make sure it is the active tab.

You can tell it is the active tab because it will be in front of the other tabs and the tab at the top of it will be a different color.

With the column selected, use your mouse and click on the Cut command in the Clipboard group

Remember from Lesson 2-1, the cut command is accessed by clicking on the small pair of scissors. When you click on the scissors the dark border around the Food column will change to a moving dashed line (like marching ants).

Move your mouse over to the "H" column and click on the letter "H"

The "H" column will now be highlighted in light blue and ready to accept the needed text.

Using your mouse, click on the Paste button in the Clipboard Group

The Paste command is in the Clipboard Group and looks like a small clipboard. As soon as you click on the Paste command, the data that was in column G is moved to column H. The result is shown in Figure 3-2.

H1		fx	Food				
	A	B	C	D	E	F	G	H
1	Month	Rent	Insurance	Electric	Phone	Gas		Food
2								
3	January	575	195	325	125	320		200
4	February	575	195	350	125	320		200
5	March	575	195	225	125	320		200
6	April	575	195	220	125	320		200
7	May	575	195	190	125	320		200
8	June	575	195	200	125	320		200
9	July	575	195	275	125	320		200
10	August	575	195	300	125	320		200
11	September	575	195	250	125	275		200
12	October	575	195	200	125	175		200
13	November	575	195	250	125	175		200
14	December	575	195	300	125	175		200
15								

Figure 3-2

Let's recap: You selected the column with the data in it, you cut the data, and you selected the column to put the data in and then pasted the data into it.

Note: You do not have to cut the entire column; you could cut a single cell or a series of cells, or even a row or series of rows.

Now we will add some data into the worksheet. It seems that we forgot to put our cable TV bill into our budget.

Select cell G1 by clicking on it

To enter information into a cell we first have to select it.

Type Cable and then press Enter

Now you have the label for the column and we can enter some values into the column.

Select cell G3 and type 125 in it and then click the Enter button on the Formula Bar

The Enter Button is the check mark on the Formula Bar. You can see this in Figure 3-3.

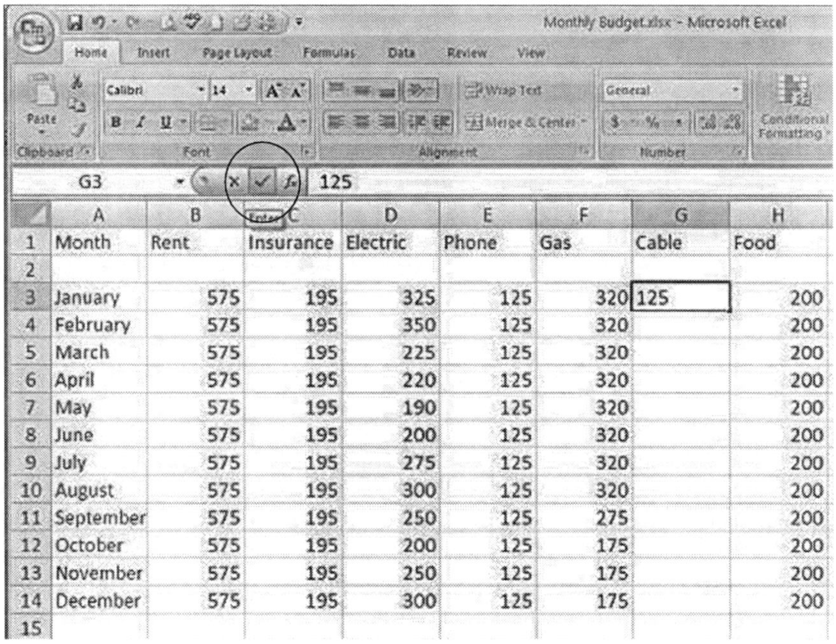

Figure 3-3

Using the Enter Button instead of actually pressing Enter on the keyboard will allow us to still have the cell G3 selected. We will not automatically drop down to cell G4. If we used the Enter key on the keyboard, we would have to go back and click on cell G3 to select it so we could copy it.

Now we have the price for our monthly cable bill in the January cell, we can copy and paste this value into the other cells.

With cell G3 still selected, click on the Copy button on the Clipboard Group

The Copy button is the one that looks like two pieces of paper side by side. It is located directly beneath the Cut button. If you have forgotten, go back and see Figure 2-2.

When you clicked the Copy button the screen will change to reflect that the selected cell has been copied to the clipboard (remember the marching ants).

Click on cell G4 and then click the Paste button on the Clipboard Group

Cell G4 now has 125 in it. The data has been successfully copied from cell G3 and put into cell G4. Notice that cell G3 still has the marching ants around it, indicating that it can be pasted again.

Repeat this process for cell G5

Now cells G3, G4, and G5 all have 125 in them. We still need to finish the rest of the year.

Select cells G6 through G14 with the mouse

You can select more than one cell by moving the mouse to the first cell that you want to select and click the left mouse button and continue to hold it down while you drag the mouse to the last cell that you want to select. Then you need to release the left mouse button to finish the select process. The result should look like Figure 3-4.

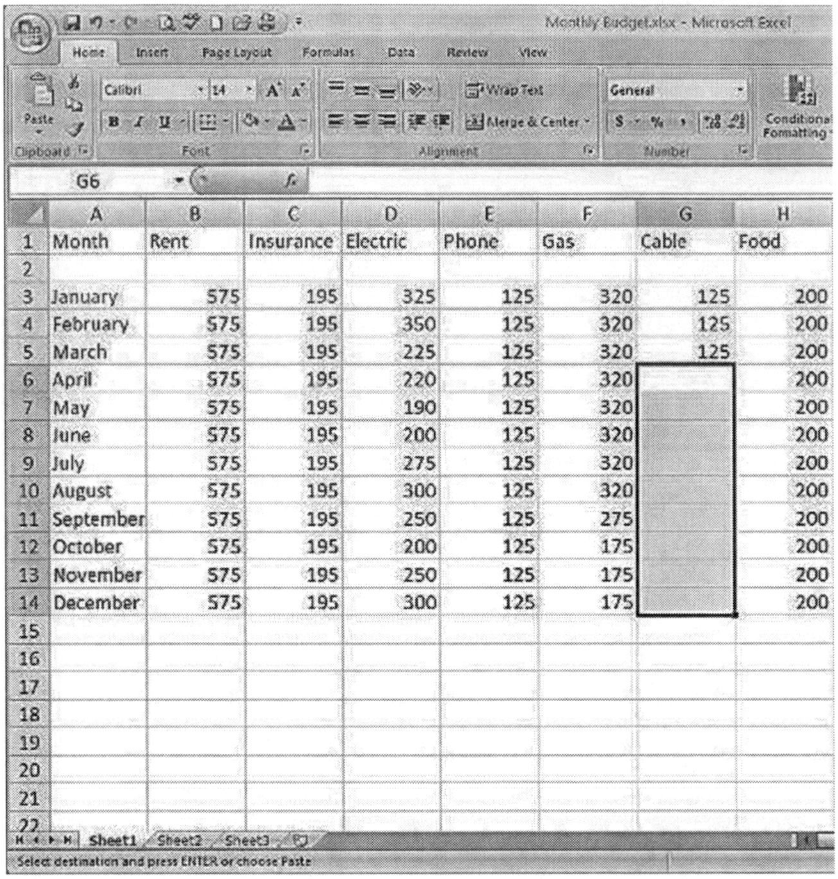

Figure 3-4

Click the Paste Button with the mouse

The rest of the column should now have 125 in each of the selected cells. Your worksheet should now look like Figure 3-5.

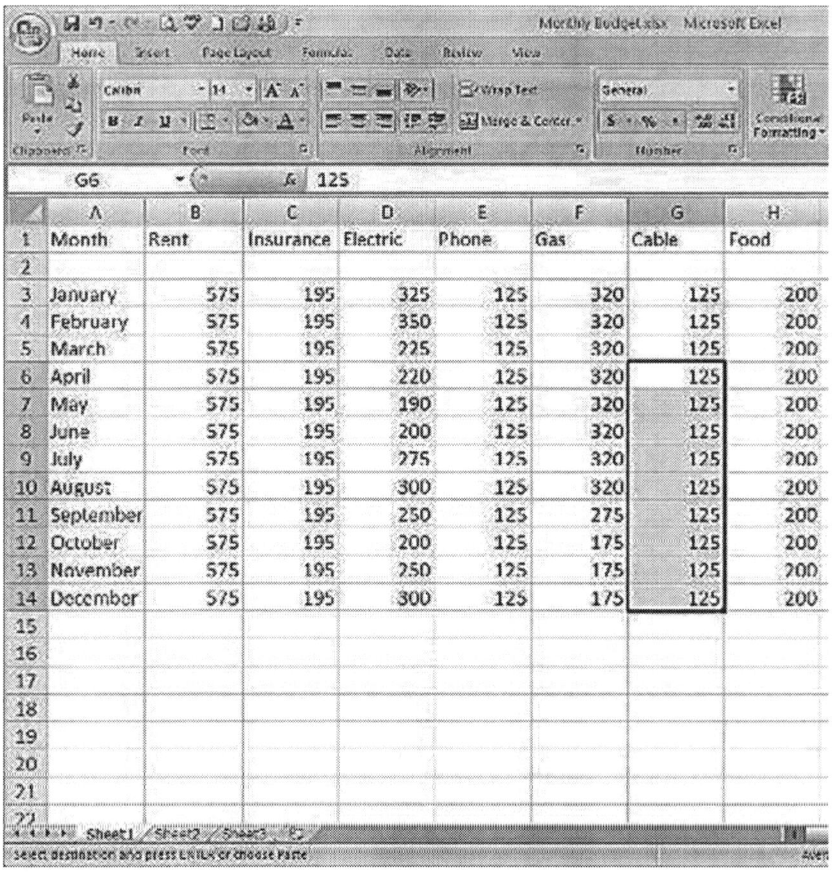

Figure 3-5

Now you have learned how to Cut, Copy, and Paste in a worksheet.

Lesson 3 – 2   Deleting Text and Values

There will come a time when you need to remove the contents of a cell. When this time comes, you will have to delete the contents of the cell. There are a few good ways to get this done and one way which will give you some results that you were not expecting. First I will cover the correct way to delete the contents of a cell and then the incorrect way.

If necessary open the Monthly Budget workbook

In our Monthly Budget example, you just remembered that you were going to be gone on vacation in June and July, and there will be no cable bill for those two months. Now we need to remove these values from our worksheet.

Click on cell G8

This cell represents the cable expense for the month of June.

Using the KEYBOARD press the Delete key

The contents of the cell are now gone. There is another way to delete the contents of a cell by using the Ribbon.

Note: The Backspace key will also delete the contents of a cell.

Note: The spacebar will also delete the contents of a cell.

Click on cell G9

This cell represents the cable bill for the month of July.

Click on the Home Tab and go down to the Editing group

Click on the Clear button

A drop down menu will appear with the following choices: Clear All, Clear Formulas, Clear Contents, and Clear Comments as shown in Figure 3-6.

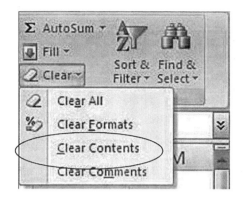

Figure 3-6

Click on Clear Contents

The content of the cell is now gone.

This is the way to delete the contents of a cell.

We will discuss the way NOT to delete the contents in Lesson 3-6.

Save your changes

Lesson 3 – 3   Using Drag and Drop

If you need to move text from one cell to another cell, you could do it in several ways. We have discussed the Cut and Paste method. Or perhaps you would prefer to delete the information in one cell and retype it in the new cell. There is an even easier way to move text: with drop and drag.

If necessary open the Monthly Budget workbook

Cells G8 and G9 should be empty, since we deleted the contents of both cells in the last lesson. If they are not, delete the contents of both of them.

Click on cell G10

This cell represents the cable bill for August. Our plans have changed and we con not go on vacation until August. Now we need to move the information from cells G10 and G11 into cells G8 and G9.

Move the mouse until the pointer changes into an arrow pointing in each direction.

This will happen when the pointer gets on one of the lines or corners. You can use any of the corners except the bottom right corner, this is for something special and we will discuss it in Chapter four. For the sake of the lesson just use the top left corner.

When the pointer changes to the arrows click the left mouse button and hold it down then Drag the mouse up until cell G8 is surrounded by a gray outline, and then release the mouse button

Cell G10 should be empty and cell G8 should have 125 in it. As you were moving the mouse you should have seen a small box moving along with you. This box shows the name of the cell that the information will go into if you let go of the mouse button. If you missed seeing this you can watch as you do the next cell.

Using the same procedure move the contents of cell G11 to cell G9

This procedure can also be used for moving multiple cells that are next to each other. If you had wanted to move both cells at once, you could have clicked on cell G10 and drug the mouse down until cell G11 was also highlighted and then followed the above procedure to move both cells at the same time.

Save your changes

Lesson 3 – 4   Using Undo and Redo

Before we continue on in our discussion of Editing, there is one other thing that will come into play: Using the Undo and Redo feature.

Being human, sometimes we make mistakes. If I had not been forced to use this feature over and over again during the writing of this book, it probably would not seem as important as it actually is. Microsoft, in their infinite wisdom, looked into the future and knew that I was going to use their product and added this feature probably just for me. I will explain it to you just in the off chance you may need it.

The Quick Access Toolbar is shown in Figure 3-7, and is the home of the Undo and Redo button.

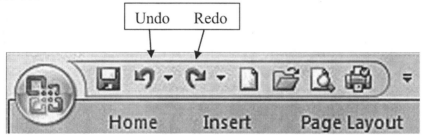

Figure 3-7

Any time you copy, type, cut, paste, or do almost anything the Undo button becomes available. If you make a mistake, you can click the Undo button and everything will be as it was before you made the mistake. Now let's be realistic here, Excel will not know that you didn't really want to do the silly thing that you just did. That means that you can't make the mistake today and tomorrow when you realize that you made the mistake, expect Excel to undo it. If you click the Undo button, Excel will undo the last thing that you did, not the mistake you made five minutes ago.

If you entered data into twenty cells and realized that the first one was incorrect and clicked the Undo button, cell twenty would be the one Excel would undo, not cell one with the mistake.

After you undo something, the Redo button will become available, so that you can redo what was originally there, or undo the undo. We will use this a little later and it will make more sense.

Lesson 3 – 5   Inserting Columns, Rows, and Cells

In Lesson 3-1 we saw how to move an entire column to make room for a new column to be added. There must be an easier way to accomplish this, and there is.

If necessary open the workbook named Monthly Budget

Select column E (Phone)

We just realized that we not only have an electric bill every month but we also have a natural gas, or propane, bill every month. We have to add this to our budget and the logical place is next to the electric bill. If you have forgotten how to select the column, you might want to go back to lesson 3-1 for a recap.

On the Home Tab inside the Cells group is a command called Insert (see Figure 3-8).

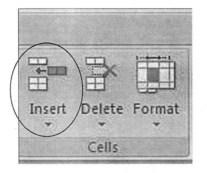

Figure 3-8

Click the down arrow on the Insert Command

This will bring the Insert menu to the screen (see Figure 3-9).

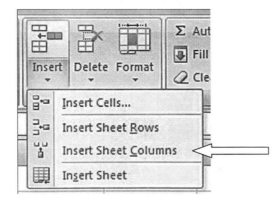

Figure 3-9

Click Insert Sheet Columns

This will insert a new column to the <u>left</u> of the highlighted column. The new column can be seen in Figure 3-10.

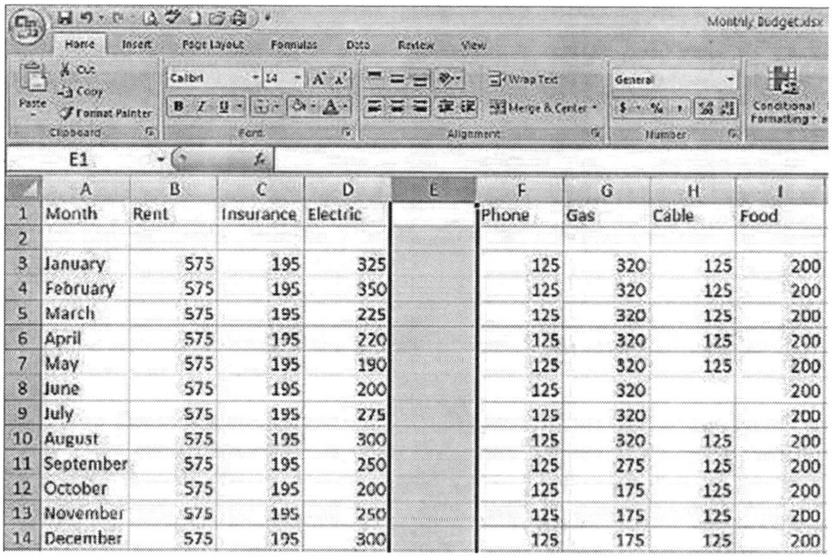

Figure 3-10

There is an even easier way to do this. After selecting the column, you could just click on the Insert button and Excel would have automatically inserted a column to the left of the selected column.

Put Propane **in for the column label (cell E1)**

Fill in each of the values as shown in Figure 3-11

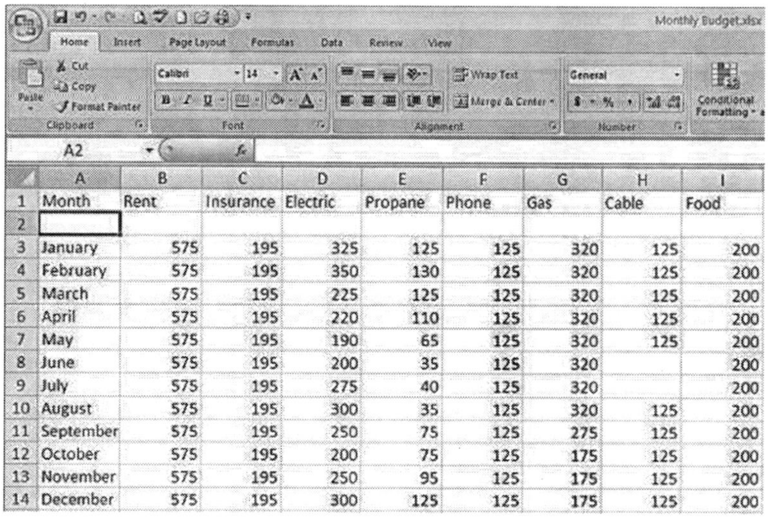

Figure 3-11

93

Save the changes to your workbook

Now that we can insert a column, what if we want to insert a new row?

Note: Do not save any of the changes that we make for the rest of this lesson.

Inserting a Row is almost identical to inserting a column. The only difference is that you select a row instead of a column before you inset.

Select the row that shows the expenses for July

This can be done by clicking the mouse on the number 9 on the left side of the worksheet.

The result of selecting the row will look like Figure 3-12

	A	B	C	D	E	F	G	H	I
1	Month	Rent	Insurance	Electric	Propane	Phone	Gas	Cable	Food
2									
3	January	575	195	325	125	125	320	125	200
4	February	575	195	350	130	125	320	125	200
5	March	575	195	225	125	125	320	125	200
6	April	575	195	220	110	125	320	125	200
7	May	575	195	190	65	125	320	125	200
8	June	575	195	200	35	125	320		200
9	July	575	195	275	40	125	320		200
10	August	575	195	300	35	125	320	125	200
11	September	575	195	250	75	125	275	125	200
12	October	575	195	200	75	125	175	125	200
13	November	575	195	250	95	125	175	125	200
14	December	575	195	300	125	125	175	125	200

Figure 3-12

Click on the small arrow on the Insert button as you did previously and select Insert Sheet Rows

A new row will be inserted directly <u>above</u> the selected row as shown in Figure 3-13.

	A	B	C	D	E	F	G	H	I
1	Month	Rent	Insurance	Electric	Propane	Phone	Gas	Cable	Food
2									
3	January	575	195	325	125	125	320	125	200
4	February	575	195	350	130	125	320	125	200
5	March	575	195	225	125	125	320	125	200
6	April	575	195	220	110	125	320	125	200
7	May	575	195	190	65	125	320	125	200
8	June	575	195	200	35	125	320		200
9									
10	July	575	195	275	40	125	320		200
11	August	575	195	300	35	125	320	125	200
12	September	575	195	250	75	125	275	125	200
13	October	575	195	200	75	125	175	125	200
14	November	575	195	250	95	125	175	125	200
15	December	575	195	300	125	125	175	125	200

Figure 3-13

94

Since the entire row was highlighted, if you had clicked the Insert Button instead of the down arrow, Excel would have automatically inserted the row for you.

Click the Undo button on the Quick Access Toolbar

The new row disappears just as though you had never inserted it. That was cool wasn't it?

What if you wanted to insert some cells and not an entire column or row?

Click the mouse on cell C6 and hold down the left mouse button

Drag the mouse down and to the right until cells C6 through I6 and C7 through I7 are highlighted and then release the mouse button (see Figure 3-14)

	A	B	C	D	E	F	G	H	I
1	Month	Rent	Insurance	Electric	Propane	Phone	Gas	Cable	Food
2									
3	January	575	195	325	125	125	320	125	200
4	February	575	195	350	130	125	320	125	200
5	March	575	195	225	125	125	320	125	200
6	April	575	195	220	110	125	320	125	200
7	May	575	195	190	65	125	320	125	200
8	June	575	195	200	35	125	320		200
9	July	575	195	275	40	125	320		200
10	August	575	195	300	35	125	320	125	200
11	September	575	195	250	75	125	275	125	200
12	October	575	195	200	75	125	175	125	200
13	November	575	195	250	95	125	175	125	200
14	December	575	195	300	125	125	175	125	200

Figure 3-14

Click the Insert Button and choose Insert Cells

Excel will insert cells into the selected area, but it needs to know what to do with the existing cells. You will get the Insert Dialog Box that is shown in Figure 3-15,

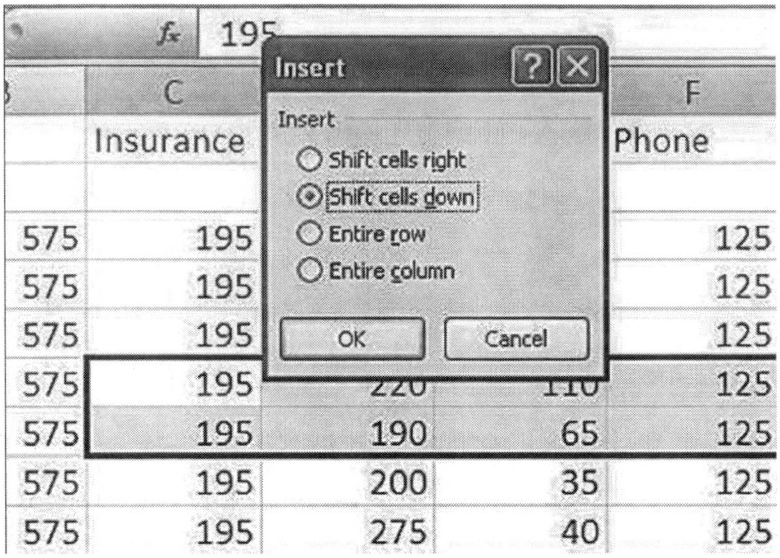

Figure 3-15

The choices are here for you to make. Do you want to shift the existing cells to the right or down? Perhaps that was not what you wanted. You could choose to insert an entire row or you could insert an entire column.

The default choice is to move the selected cells down. If you had simply clicked the Insert button instead of the down arrow, this is what Excel would have done.

Click OK and see what happens

You can see that all of the cells were shifted down. This is not what we want for our budget. Do you remember how to rid of this mistake?

Click the Undo button

Boy that looks better.

Repeat the above steps and choose Entire Column from the Insert Dialog Box

You will notice that enough columns were inserted to cover the cells that were selected. Seven columns were inserted because the selected cells were covering seven columns. Since we don't want this in our budget, make sure you click the undo button when you are finished.

Lesson 3 – 6   Deleting Columns, Rows, and Cells

In the last lesson we learned how to insert columns, rows, and cells. In this lesson we will learn how to delete columns, rows, and cells. Deleting columns, rows, and cells is almost identical to inserting them.

First we will delete a column. There are basically two ways, each with its own advantages.

If necessary open Monthly Budget

Click on Column E that we inserted in the last lesson.

The entire column will be highlighted and ready for us to do something with it.

Press the Delete button on the keyboard

All of the data that was in the column is now gone as shown in Figure 3-16.

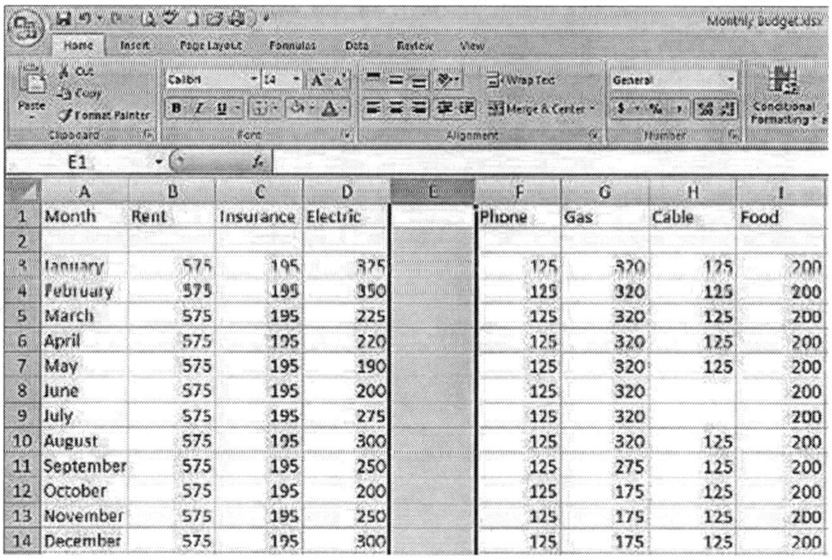

Figure 3-16

This can good or bad depending on what you wanted to accomplish. If you wanted to replace the existing data, this would be good. If on the other hand, you actually wanted to delete the column, removing the data would not have helped you. As you can see the column is still there.

Click the Undo button to restore the contents

97

Now that the column is back to where you started, let's try it again.

Click the down arrow on the Delete command in the cell group of the Home tab

You will get the Delete Dialog box as shown in Figure 1-17.

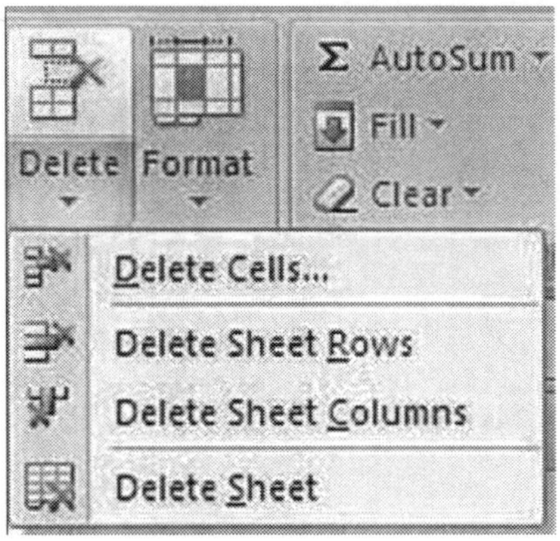

Figure 3-17

Click on the Delete Sheet Columns choice

Now the column is actually deleted, not just the contents but the whole column. It is as if the column was never there. Figure 3-18 shows the result of clicking the Delete Sheet Columns choice.

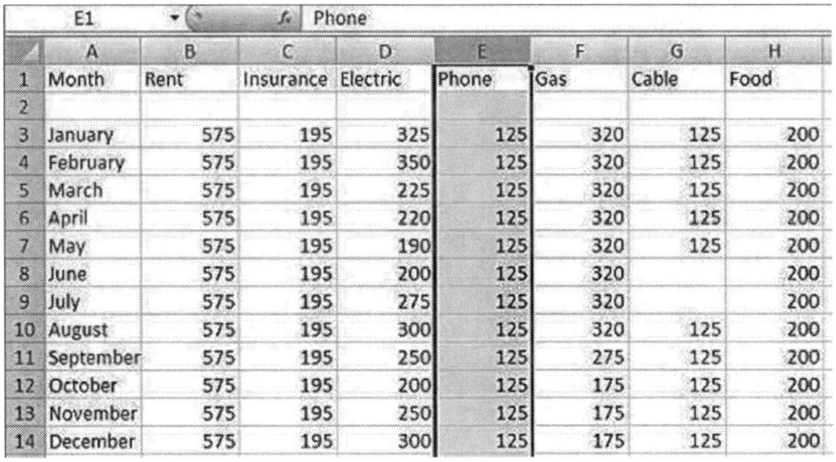

	A	B	C	D	E	F	G	H
1	Month	Rent	Insurance	Electric	Phone	Gas	Cable	Food
2								
3	January	575	195	325	125	320	125	200
4	February	575	195	350	125	320	125	200
5	March	575	195	225	125	320	125	200
6	April	575	195	220	125	320	125	200
7	May	575	195	190	125	320	125	200
8	June	575	195	200	125	320		200
9	July	575	195	275	125	320		200
10	August	575	195	300	125	320	125	200
11	September	575	195	250	125	275	125	200
12	October	575	195	200	125	175	125	200
13	November	575	195	250	125	175	125	200
14	December	575	195	300	125	175	125	200

E1   *f* Phone

Figure 3-18

Note: This could have also been accomplished by clicking the Delete button, instead of clicking the drop-down arrow and then choosing to delete the column.

Before we look at deleting rows, click the Undo button on the Quick Access Toolbar

Now let's look at deleting Rows.

Select Row 7

This can be selected by clicking on the number 7 on the left side. By now you know that if you press the Delete key on the keyboard the only thing that will go away is the information, not the row. To delete the row we must use the Delete command of the Cells group on the Home Tab of the Ribbon.

Click on the Delete button

The entire row containing the information for the month of May is now gone and the remaining rows have moved up to fill in the blank row.

Click the Undo button

Now everything goes back to the way it was before the delete.

Now let's look at deleting an individual cell.

Click on cell E 8

We want to delete this cell from our worksheet.

Click on the down-arrow of the delete command and choose Delete Cells

Now you get to decide what to do with the remaining cells. The dialog box that comes to the screen is shown in Figure 3-19.

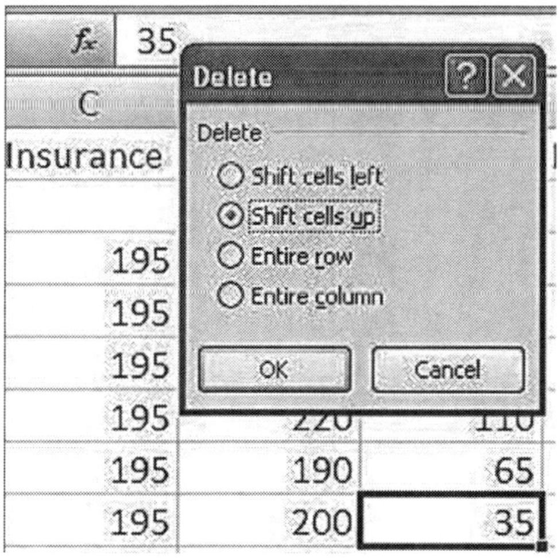

Figure 3-19

You can move the remaining cells up or move the cells from column beside it over to fill the empty place. The default choice is to move the remaining cells up to fill the empty space created by deleting the cell.

Click the OK button and watch what happens to the worksheet

All of the remaining cells move up one position.

Note: If you click on the Delete Button, not the down-arrow, that cell will be deleted, not just the contents the entire cell will be deleted. The result is that all of the other cells will move up one position so that the deleted cell is replaced by the one below it and all of the others will also move up to replace the one above it.

Click the Undo button

Our worksheet is now back to the way it should be.

Remember when I told you that I would show you the incorrect way to delete information from a cell? Well this is it. If you only want to remove the data that is in a cell, DO NOT use the delete button. The cells will be removed along with the data that was in them.

Lesson 3 – 7   Finding and Replacing Text

There may be times when you need to find a particular piece of information. Excel has included a Find feature just for that. It would be advantageous for you to make this worksheet yourself, but it is long so I have provided it for you on the EZ 2 Understand Computer Books website. **Toward the beginning of the book is a page that has Important Notice on it. This will tell you how to download the files from the website.**

Open the workbook entitled Trips from the downloaded files

This spreadsheet is a fictitious set of data entries concerning several salespersons and trips they have booked for the company. In this lesson we will see how to find all entries that were booked by a salesperson and all trips to the same place regardless of the salesperson.

The first task we want to perform is to find a specific salesperson.

On the Home Tab and under the Editing Group, click on the Find command

This will bring a menu to the screen as shown in Figure 3-20.

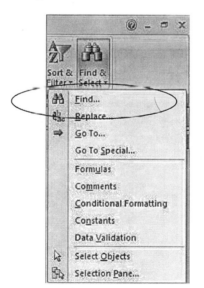

Figure 3-20

Click on the Find choice

> This will bring the Find and Replace Dialog Box to the screen as shown in Figure 3-21.

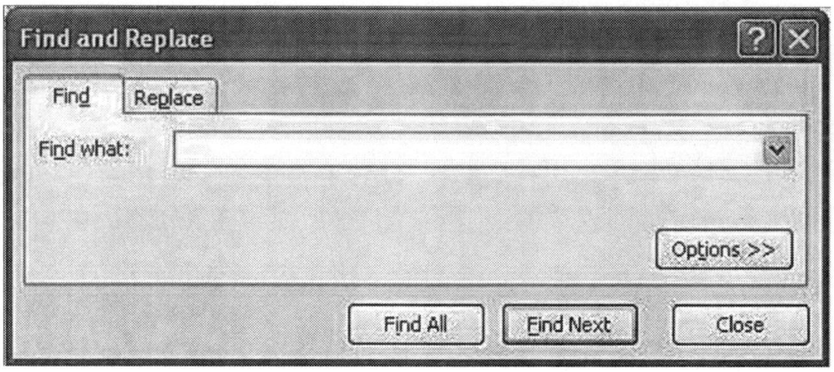

Figure 3-21

> Now we can type in the text that we want Excel to search for, and if Excel finds any text that matches it will enclose it in an outline.

Type William **and click Find All**

> When you click the Find All button, a new dialog box will appear on the screen. This can be seen in Figure 3-22.

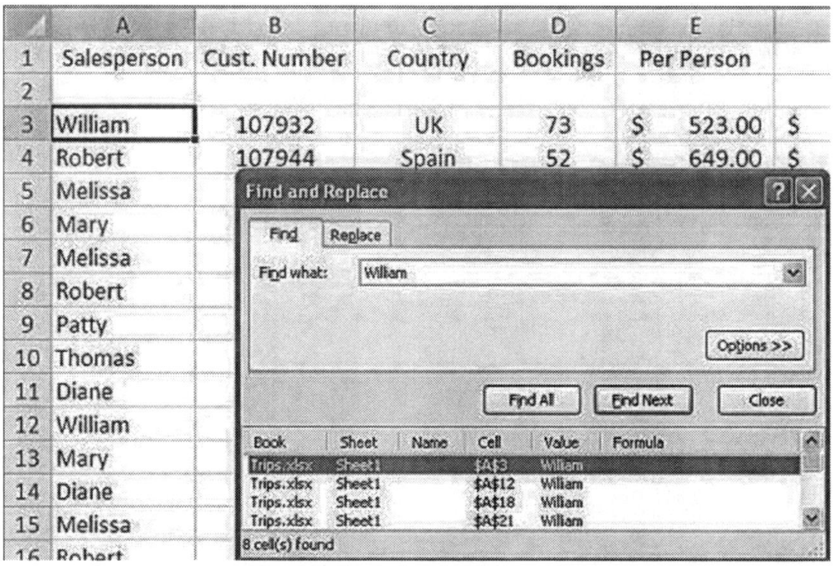

Figure 3-22

102

As you can see, the first time the Find feature encountered the typed text, it selected that cell. You can tell this by the black outline around the cell A3. Also you should note that the bottom section of the dialog box shows all of the occurrences of the text William.

Click the Find Next button

The outlined cell has now changed to A12 because that is the next occurrence of the text William. Let's try it again only with a value this time.

In the Find what textbox type 523 and then click the Find All button

If you click the Find All button the search will start at the first cell (A1). If you had typed 523 and pressed the Enter button, the search would not have started at cell A1; it would have started from the active cell and not found any cell before the active cell.

Look at the bottom section again, the part that shows every occurrence of the text. The first one in the list should be highlighted in blue.

Click on the second listed occurrence

Notice that as you move the mouse over each listing, a blue line is visible under each listing as the mouse passes over it. This shows you which listing you will go to if you click the mouse.

Now we will pretend that one of our employees has left the firm and you are going to assign all of his customers to a new employee. You decide to just change the name of the salesperson from William to James.

Click on the Replace tab of the Find and Replace Dialog Box

The Replace tab jumps to the front as shown in Figure 3-23.

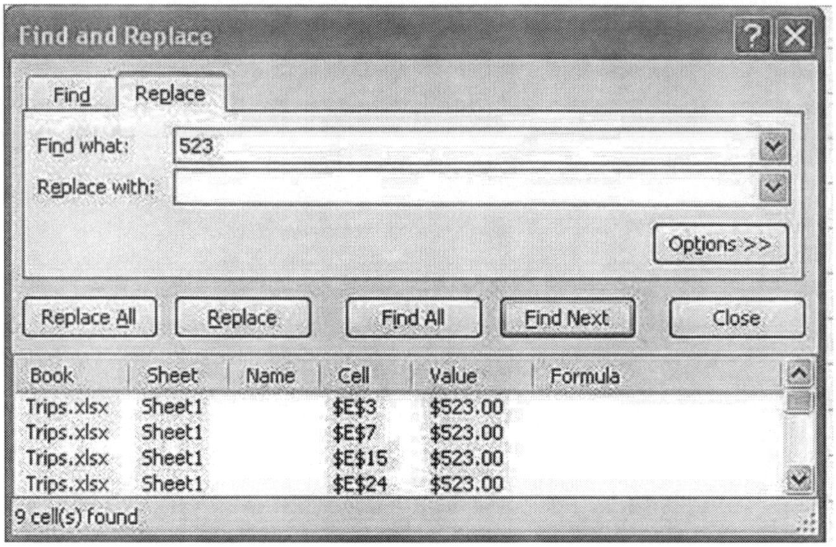

Figure 3-23

In the Find What textbox type William and in the Replace With textbox type James

Now it gets a little tricky here. If you click on the Replace button and the active cell does not have William in it, you will get a Message box stating that Microsoft Excel cannot find a match. This is because the Replace command only looks at the active cell (the one with the outline around it).

If you click the Find Next button, Excel will start looking from the active cell and not start at the beginning and search from there.

If you click the Find All button, Excel will start the search from the beginning and locate every occurrence of the Find What text. The active cell will change to the first occurrence of the Find what text, and then you can use the replace command to replace that occurrence of the text.

If you click on the Replace All button, every occurrence of the find text will be replaced by the text that is in the Replace With text. **Be careful with this.** It will replace all occurrences of the find text with the replace text. **Make sure this is what you want to do before you click the replace all button.**

If you did click the Replace All button, click the Undo button

Save this spreadsheet in the My Documents folder (Use the name Trips)

Click the Close button

Lesson 3 – 8   Sorting Data

There will be times when it is easier to find data if it were in alphabetical or numerical order, not just in the spreadsheet. Hey, Microsoft thought of that too. In this lesson, we will explore this feature.

Open a new workbook

In this workbook we will create a list of our customers with addresses and phone numbers.

Type the following into your new workbook.

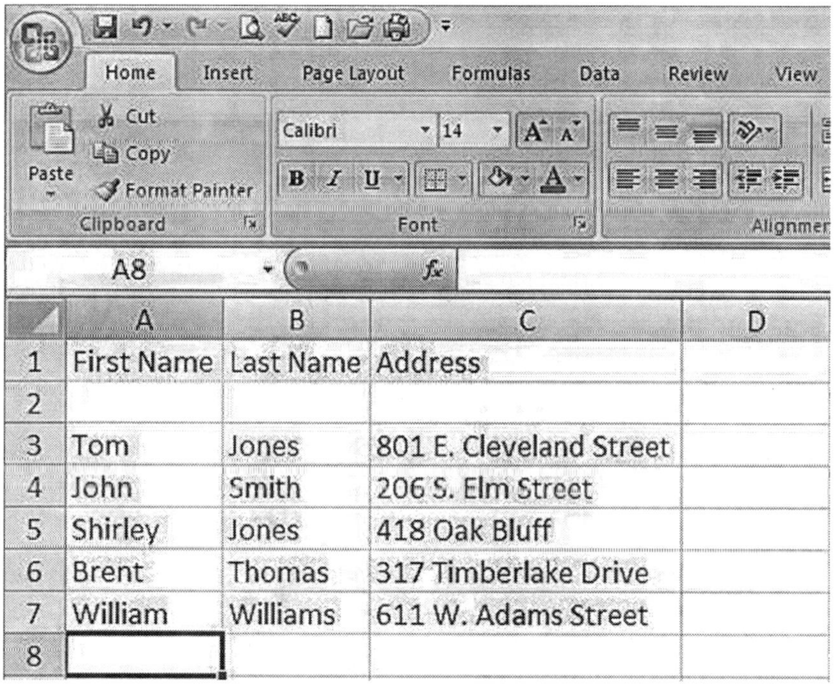

Figure 3-24

The first thing you will probably notice is that column C in our example does not look like your column C. My column C automatically adjusted to allow the text to fit into it. I will show you how to make any column automatically adjust to the size of the text.

Click on the letter C to select the column

The column will now be highlighted in blue.

On the Home Tab and in the Cells Group, click on the down-arrow on the Format command

The following should jump onto your screen.

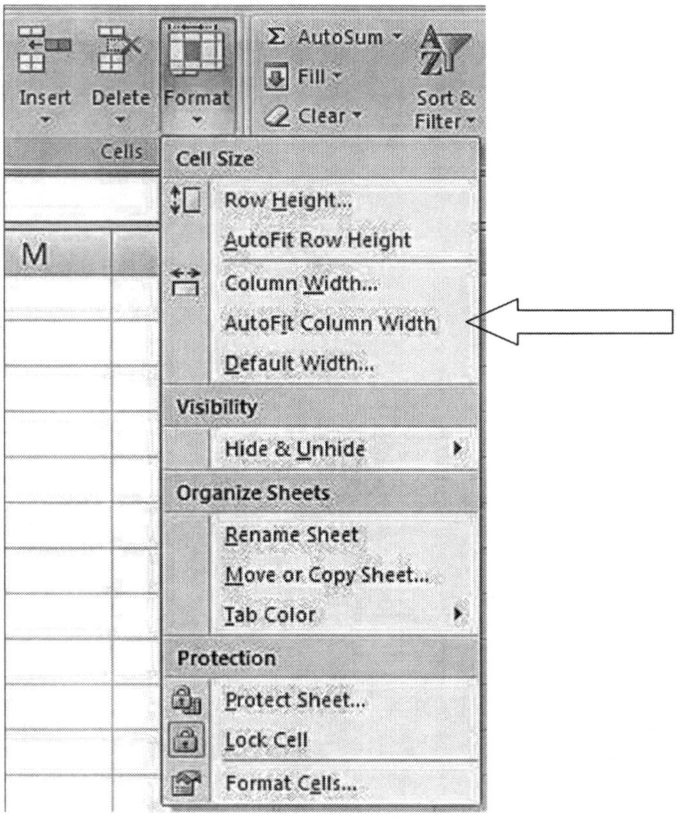

Figure 3-25

Click the AutoFit Column Width choice

The column will automatically adjust itself so the text will fit into it. Now your worksheet will look like mine.

There are a few things that you will need to get use to, if you are going to sort the data in your worksheet. One of these things is what happens to the spaces (cells with nothing in them). Just for fun let's sort the data in column A.

First we will use the Sort that is on the Home tab.

Click on column A

With the column selected we can sort the data that is in it.

We will be using the Sort and Filter that is in the Editing Group (see Figure 3-26).

Figure 3-26

Click on the Sort and Filter button and the drop-down menu seen in Figure 3-27 will appear.

Figure 3-27

For our practice sort, let's use the A to Z sort.

Click on Sort A to Z

Guess what? Excel doesn't know what to do with the rest of the cells. Excel knows that there are more cells to the right of the selected cells, but it doesn't know what to do with them. Let's take a look at the warning that Excel is giving to us (see Figure 3-28).

107

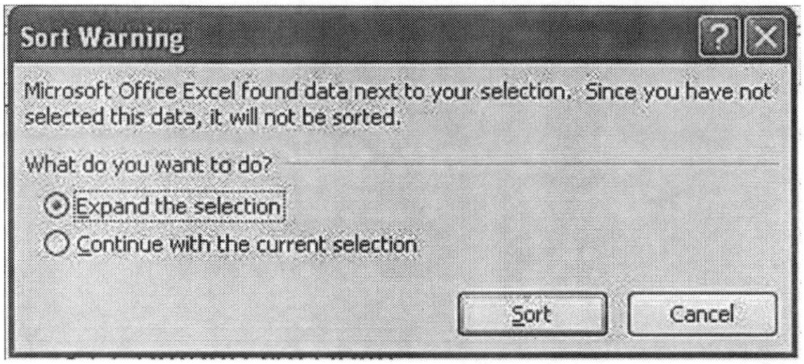

Figure 3-28

In the bottom part expand the selection is chosen. If you click on the Sort button, all of the data will be sorted because you chose to expand the selection to the other columns. If you had clicked the radio button next to continue with current selection, only the first column would have been sorted. Excel tries to do what it thinks is best for the situation and it assumes that the entire worksheet should be sorted. That is why this is the default choice.

Click the Sort button and see what happens

The result of the sort is shown in Figure 3-29.

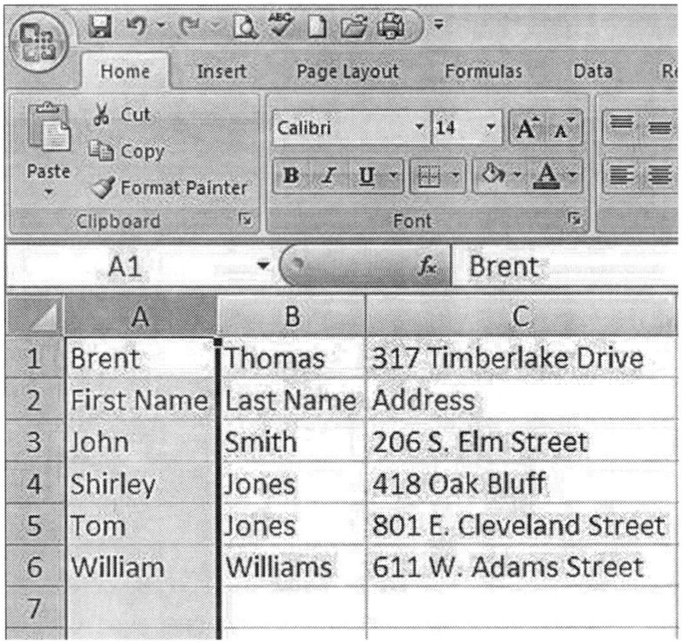

Figure 3-29

In Figure 3-24 you had seven rows in use, with row two being blank. In Figure 3-29 the first six rows have data in them. The blank row was moved to row seven. I don't think that this was the result we wanted.

Click the Undo button to put our spreadsheet back the way it was before the sort

Now let's try to sort only the data not the labels on the top row.

Click on cell A3 and drag the mouse until cell C7 is also selected

Your screen should look like Figure 3-30.

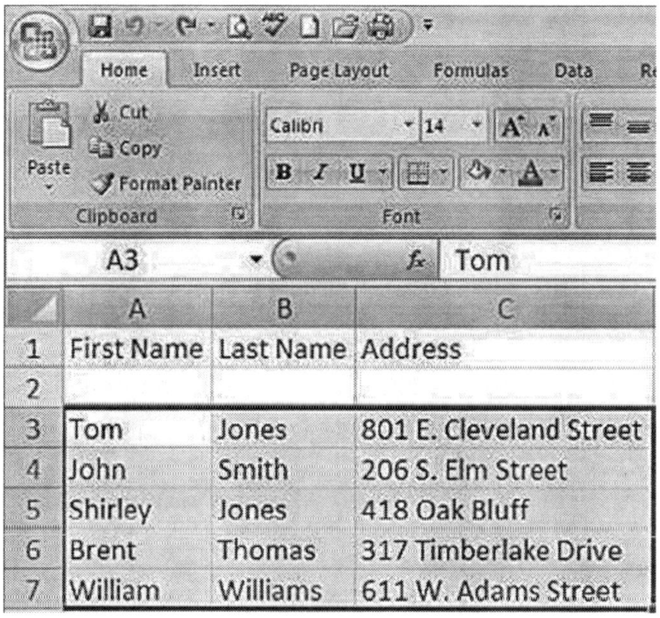

Figure 3-30

Click on the Sort and Filter button and choose Sort A to Z

The result should look like Figure 3-31. All of the data is sorted alphabetically using the first name as the key to sort from.

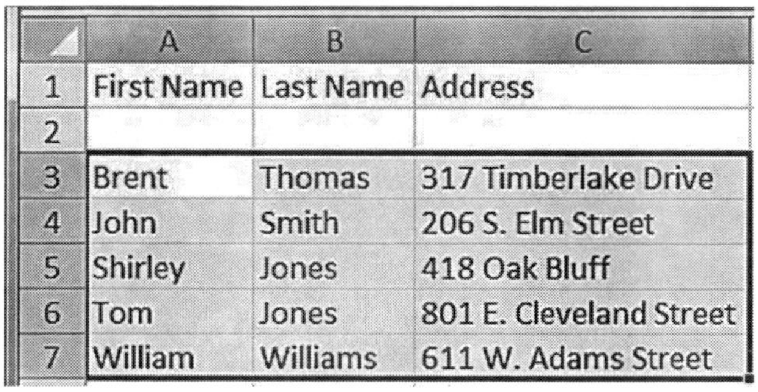

Figure 3-31

This time let's sort the data using the sort on the Data Tab.

Click the Undo button so that the name will not be in alphabetical order

Click on the Data Tab of the Ribbon

With the text still highlighted (if it is not highlighted reselect the data);

Click on the Sort A to Z with the down-arrow (the one in Figure 3-32 with the arrow)

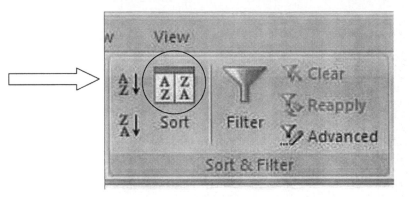

Figure 3-32

The result will be the same as it was using the other method. This method is a lot faster than using the one on the Home Tab.

Note: The data can be sorted in reverse alphabet by choosing the Z to A sort.

Now let's try something a little different.

Click the Undo button

With the data back to the way it was before the sort and still highlighted, let's sort by the Last Name instead of the First Name.

Click on the Sort button (the one with the circle around it).

A new dialog box comes to the screen as shown in Figure 3-33.

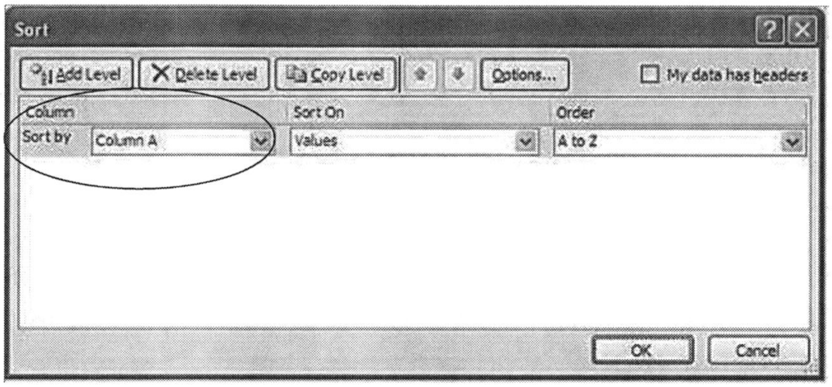

Figure 3-33

Click on the down-arrow on the column A choice

On the drop down menu choose Column B (as shown in Figure 3-34)

Click the OK button

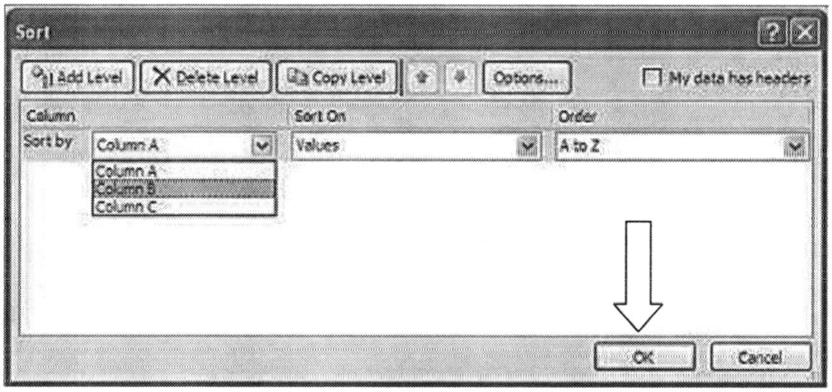

Figure 3-34

The data is now sorted by Last Name as seen in Figure 3-35.

	A	B	C
1	First Name	Last Name	Address
2			
3	Shirley	Jones	418 Oak Bluff
4	Tom	Jones	801 E. Cleveland Street
5	John	Smith	206 S. Elm Street
6	Brent	Thomas	317 Timberlake Drive
7	William	Williams	611 W. Adams Street

Figure 3-35

Save the workbook in My Documents using the name Addresses

Close the Workbook

Now we will look at using a filter to sort the data. This will work better with a larger spreadsheet, so we will use the Trips workbook.

Open the workbook named Trips that you saved in My Documents

With the larger workbook open we can pull up all of the data for a specific salesperson.

Select the column with the salesperson's name in it

Now we can set the criteria for our sort. In this sort we only want to see the sales data for Robert. We don't need to see any other data, so we will have to use the filter.

On the Data tab and in the Sort and Filter group, click on the filter command (the one with the funnel)

The screen will change only slightly. Look closely at the top of column where it says salesperson. There is now an arrow on the right side. This is shown in Figure 3-36.

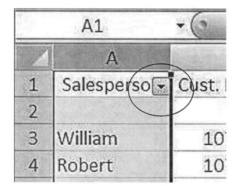

Figure 3-36

Click the arrow next to salesperson

The result of the mouse click is shown in Figure 3-37.

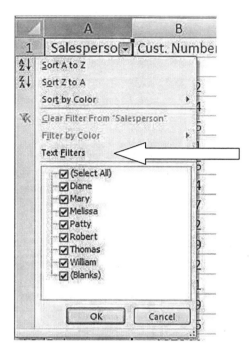

Figure 3-37

There are two ways to do what we want to do, so let's do the hard one first. We will select all of the names available, and set the filters manually.

Click on Text Filters

As you might expect, we get another menu to work with. It is shown in Figure 3-38.

Figure 3-38

Click on Equals

This will allow us to put the text that we want to sort by into the textbox ourselves. We only want to see the sales for the salesperson named Robert, so we need to put Robert in the textbox. Look at Figure 3-39 to see the dialog box where we will enter the text.

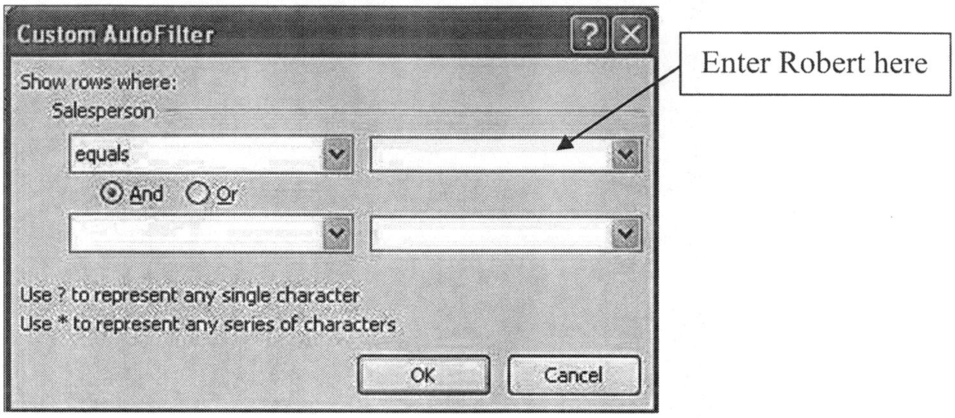

Figure 3-39

Type Robert in the textbox and then press the OK button

The result should be as shown in Figure 3-40.

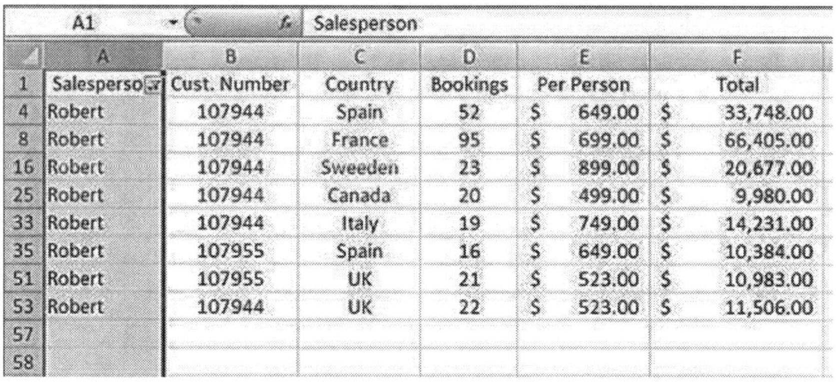

A1 | Salesperson

	A	B	C	D	E	F
1	Salesperso	Cust. Number	Country	Bookings	Per Person	Total
4	Robert	107944	Spain	52	$ 649.00	$ 33,748.00
8	Robert	107944	France	95	$ 699.00	$ 66,405.00
16	Robert	107944	Sweeden	23	$ 899.00	$ 20,677.00
25	Robert	107944	Canada	20	$ 499.00	$ 9,980.00
33	Robert	107944	Italy	19	$ 749.00	$ 14,231.00
35	Robert	107955	Spain	16	$ 649.00	$ 10,384.00
51	Robert	107955	UK	21	$ 523.00	$ 10,983.00
53	Robert	107944	UK	22	$ 523.00	$ 11,506.00
57						
58						

Figure 3-40

The only data that is shown is the data that met the criteria that we set with the filter.

To remove the filter, and restore all of the data click on the filter button a second time

As I said before there is also an easy way to do this.

Click on the down arrow again and this time click the checkbox next to Select All

Now all of the checkboxes are empty and we can select the criteria for the filter right here.

Click on the checkbox next to Robert and then click OK

You should get the same results as before with only Robert's data showing.

To restore everything, click the filter button on the Ribbon a second time

Close the workbook without saving the changes

Lesson 3 – 9   Using Spell Check

There is probably nothing more embarrassing than to have someone come up to you and correct your spelling and/or grammar in a document. Microsoft thought that this would be embarrassing for you as well and provided you with a spell checker feature. Before you let anyone see your work, it is probably a good idea to run the spell checker.

Open the workbook named Recipes that is in the downloaded files from the website

There are several typing errors in this workbook. There are more than likely some recipe errors also, so you don't have to send a list of them to me.

Click on cell A1

We want to make sure the spell checker starts from the first cell. If we had clicked on cell A3, the spell checker would have missed the first two rows on its first pass through.

Click the Spell Check button on the Quick Access Toolbar

The Spell Check button is shown in Figure 3-41.

Figure 3-41

The Spell Check Dialog Box should pop onto the screen (see Figure 3-42).

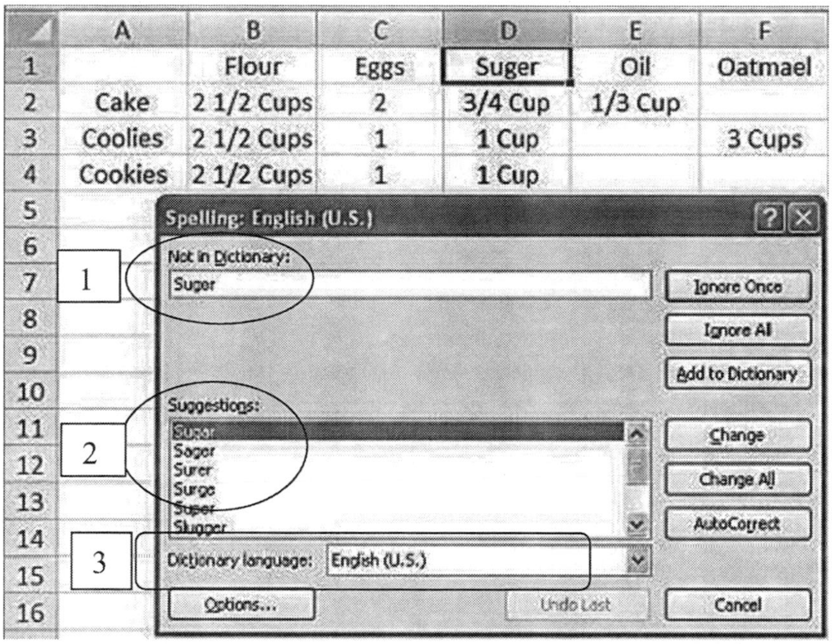

Figure 3-42

The first misspelled word should be suger. There are several things that we should take note of:

1) The word suger is not in the dictionary.
2) Directly below this there are several suggestions for us to choose from.
3) The language we are using for the dictionary is English.
4) The buttons on the right are for us to click on after we have made our choice.

Now let's look at our choices on the right.

1) We can ignore this spelling error once.
2) We can ignore all occurrences of this spelling error.
3) We can add this word to the dictionary.
4) We can change this word for the word that is highlighted under suggestions.
5) We can change all occurrences of this word with the highlighted word.
6) We can use the AutoCorrect feature to make the choices for us.
7) We can cancel the spell check.

The first normal question is why would we want to ignore a misspelled word? The most common answer is that some companies intentionally misspell word to use as part of a logo or even in their name.

If you are sure the word is spelled the way it should be spelled and it is not in the dictionary, we might want to add the new word to the dictionary. As an example, I have added my e-mail user name to the dictionary because I use it a lot and I get tired of it showing up as misspelled.

Using the AutoCorrect feature will let Excel decide which choice to use, even if there is more than one choice to pick from. If you click on AutoCorrect, Excel will replace the misspelled word with the highlighted word.

In this situation, the word is really misspelled and we need to change it with the correct spelling; sugar.

Click the Change button

The misspelled word is now replaced with the correct spelling, and the next word that Excel thinks is misspelled is brought into question.

Continue with the Spell Check until you get to the Table spoon abbreviation

As you can see the Spell Checker has no idea what this word is suppose to be, so it gives you several things that night work. None of the choices that Excel has thought of are correct, so what do we do? If this is the correct abbreviation, we could add it to the dictionary. I am not sure that there is a correct abbreviation for Table spoon, so let's get out of the Spell Check and fix the word manually.

Click on the Cancel button

Click on the Formula bar (see Figure 3-43)

Click here

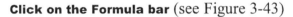

	A	B	C	D	E
1		Flour	Eggs	Sugar	Oil
2	Cake	2 1/2 Cups	2	3/4 Cup	1/3 Cup
3	Coolies	2 1/2 Cups	1	1 Cup	
4	Cookies	2 1/2 Cups	1	1 Cup	
5					

K3   x ✓ ƒx   1 Tspn

Figure 3-43

118

Edit the text in the cell to read T Spoon as shown in Figure 3-44 and then press Enter

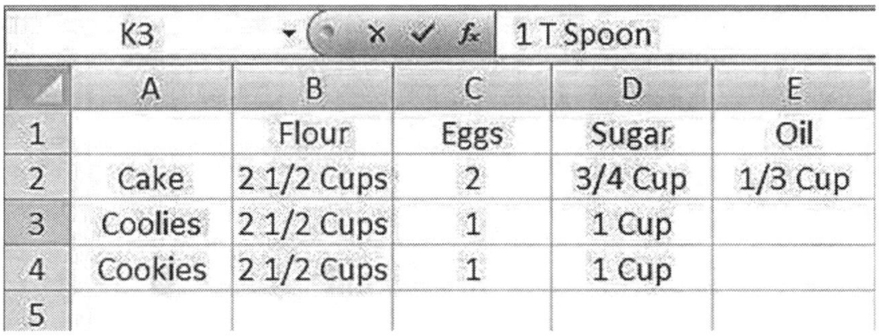

	A	B	C	D	E
		Flour	Eggs	Sugar	Oil
1					
2	Cake	2 1/2 Cups	2	3/4 Cup	1/3 Cup
3	Coolies	2 1/2 Cups	1	1 Cup	
4	Cookies	2 1/2 Cups	1	1 Cup	
5					

K3 — 1 T Spoon

Figure 3-44

Since we all know that a capital T represents a Table Spoon and a small t represents a Tea Spoon, this should be Okay and pass the spell check.

Click on cell A1

Click on the Spell Check button again

The spell checker should stop on cell K4. There will not be a choice to fix this mistake. It should also be 1 Table Spoon.

Repeat the above steps to correct the text in the cell

Run the Spell Check again to see if there are any other misspelled words

This is the last cell in the workbook, so Excel will want to know if you want it to start at the beginning of the sheet. When it asks, click Yes. Click the OK button when the spell check is finished.

Save your changes and save the workbook in My Documents

119

Lesson 3 – 10   Adding Comments

It would be nice to be able to add something extra in a cell to remind you of why it is there or something similar to that. Microsoft allows us to put a comment inside a cell and have it hidden until we want to use it.

Open the Trips workbook

I want to call your attention to cell A1, in particular the top right corner of the cell. You will notice a small red triangle in the corner. This triangle indicates that there is a comment hidden in the cell. To see the comment we need to move the mouse pointer over to the triangle.

Move your mouse pointer to the triangle and let it hover over it

You will be able to see the comment (See Figure 3-45).

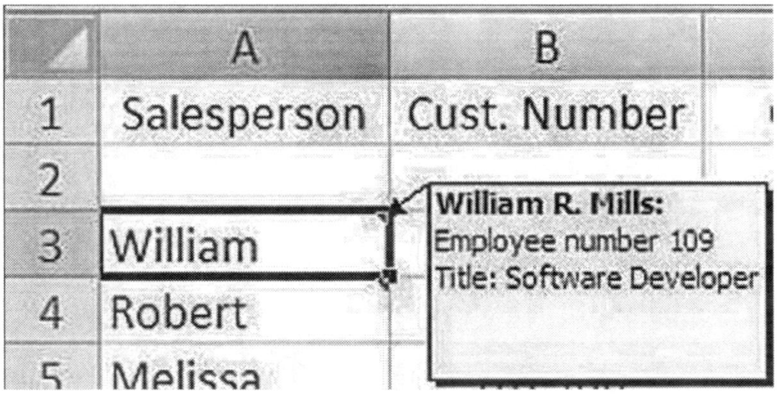

Figure 3-45

Now let's see how to add a comment to a cell.

Click on cell A4

First we have to make the cell that is going to hold the comment active. This is done by clicking on it. Now we need to get to the comment section of the Ribbon.

Click on the Review Tab

Comments are added by clicking on the New Comment command in the Comments Group. This is shown in Figure 3-46.

Figure 3-46

Click on the New Comment command

The first thing that will notice is that your name is at the top of the comment box. This may not be what you want to show. It was intended so that you can see who put the comment in the cell. I recommend that you leave it there, but you don't have to, it is up to you.

In the comment box type Robert Smith and then press Enter to go to the next line down. Finish the comment as seen in Figure 3-47

Figure 3 47

You can move from one comment to the next by clicking on the next and previous buttons. You can also make all comments visible by clicking on the Show All Comments button.

The Show/Hide Comments feature is only available if the active cell has a comment; otherwise the button is faded out so that you cannot click on it.

If the active cell has a comment and you want to delete it, you can click on the Delete button. By the way, you won't be asked to confirm the deletion the comment will just be deleted.

Close the workbook and save the changes

121

Chapter Three   Review

Cut, Copy, and Paste are used to move data in a worksheet. Cut removes the selected data and sends it to the Clipboard. Copy sends a copy of the selected text to the Clipboard while leaving the original text in place. Paste copies from the Clipboard and sends the data to the selected cell. You can select more than one cell and paste the data into all of the cells at one time.

You can delete the text in a selected cell by pressing the Delete key on the keyboard or by using the Clear button in the Editing Group which is on the Home Tab.

Drag and Drop allows you to move text and/or values without using Cut, Copy, and Paste.

Undo will allow you to undo the last thing that you did on the worksheet, unless you have closed the workbook or the Excel program.

Redo will put the workbook back the way it was before the undo.

Columns, Rows, and Cells can be inserted into and deleted from the worksheet. If you are working with cells, you must tell Excel what to do with the surrounding cells.

You can search for and find text in the worksheet. You can also have Excel replace certain text for you automatically.

Data can be sorted from smallest to largest or largest to smallest. You can also choose which column to use when sorting data.

You can use filters to only show certain data.

Comments can be added to cells in a worksheet.

Remember to spell check on all worksheets.

Chapter Three   Quiz

1) Explain the difference between cut and copy in reference to moving data.
2) Text can be pasted to only one cell at a time. **True or False**
3) Clicking the delete button that is in the Cell Group of the Home Tab will only remove the data that is inside the selected cell. **True or False**
4) Explain how to use the Drag and Drop feature.
5) Clicking the Undo button will cause Excel to remove any mistake it finds. **True or False**
6) If a column is selected, clicking the Insert button in the Cells Group will insert a column to the left of the selected column. **True or False**
7) Pressing the Delete key on the keyboard will delete the text inside the selected cell and remove the cell from the worksheet. **True or False**
8) The Replace All button in the Find and Replace Dialog Box will ask for confirmation at every occurrence of the word it is searching for. **True or False**
9) When sorting data, empty cells are ignored and left in place in the worksheet. **True or False**
10) When Spell Check finds a word that is not in the dictionary, explain how you add the word to the dictionary.

Chapter Four — Mathematical Calculations and Formulas

Now that we have gotten the basics out of the way, we can start using Excel the way it was intended to be used. We can start doing mathematical calculations and entering formulas.

This is the real power of Excel. Excel can perform long and complicated mathematical calculations. Remember this is a beginner's guide. We will not be going into the world of complicated mathematical calculations such as trigonometry or engineering calculations. We are going to keep our focus on regular things that you will be using every day in Excel, nothing to complicated.

In this chapter we are going to stick with the basic math (additions, subtraction, multiplication, and division). There will be a few other things to make your life a little easier, but nothing too bad.

Lesson 4 – 1   More on Cell Names

Sometimes remembering a cell name and its reference is hard to do. For example, if you are working on a large spreadsheet and you reference cells D9 through D13 to get a total or an average, if you see this reference later you may not remember what it was used for. It would be a lot easier if you gave this reference a name such as Feb_Total.

In this lesson we will practice giving names to cells and also a range of cells. We will be using a workbook with the weekly attendance for Sunday morning worship services.

Open the workbook titled Church Attendance

This lesson is included with the files that you downloaded.

Select the cells B3 through B7 and then click in the name box that is in the Formula bar

If you are unsure how to do this, you select a range of cells by clicking the mouse on the first cell and hold the mouse button down as you drag the mouse to the last cell and then release the left mouse button. Your screen should look like Figure 4-1.

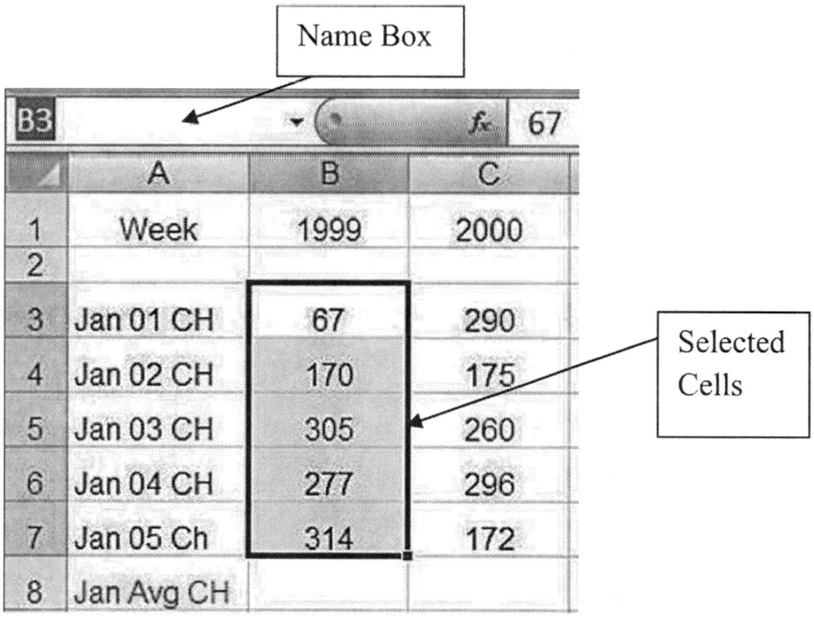

Figure 4-1

In the Name Box type Jan_99

You will need to put the underscore in the name. Excel does not like spaces in a name and you will get an error message box if you try it. Jan99 is also a reserved name and cannot be used.

Now when we reference these cells we can type Jan_99 instead of B3:B7.

Click on cell B8 and type the following =Sum(Jan_99) / 5 and then press Enter

This formula adds the numbers in the range we set up in the last step and then divides that number by 5. We will dive deeper into the formulas in rest of this chapter.

This could also have been written like this: = Sum(B3:B7) / 5.

It seems to me that typing Jan_99 is easier than typing B3:B7 and it is a lot easier to remember.

Using this same procedure, enter a name for the remaining months of 1999

In case you need a little reminder, start with cells B9 through B13 and give it the name Feb_99. Just to make life easier (for me) only use the first three letters of the month in the name. That is how I am going to refer to them.

Want to see something really cool? Try this out.

Click on cell B14

This is where we are going to enter the average attendance for the month of Feb.

Type the following = Sum(

Click on the Formula Tab and then on the "Use in Formula" command in the Defined Names Group

This is shown in Figure 4-2.

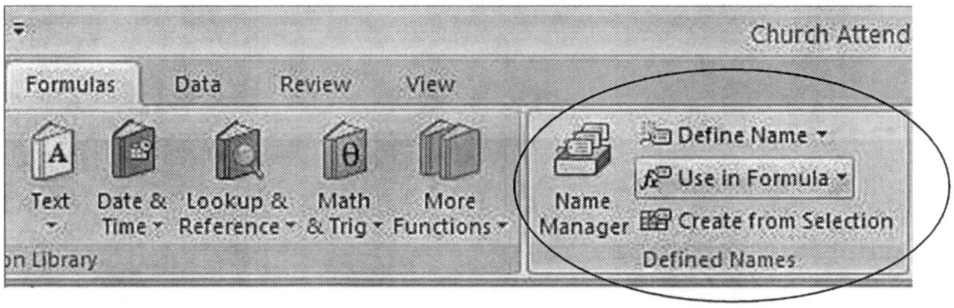

Figure 4-2

126

Click on the name Feb_99 and then finish the formula so it reads = Sum(Feb_99) /4

We divided by four instead of five because we only had data for four Sundays. In January we have data for five Sundays, so we had to divide by five.

Finish the rest of the averages for the year 1999

Keep a close eye on the number of weeks in the month and divide only by the number of Sundays you have data for.

A name does not have to be only for a range of cells, it can be for one cell only.

Click on cell B8 and give it the name Jan_99_Avg

If we finished the averages for the workbook there would be 96 averages. Trying to remember all of the cells would be difficult. Referencing Jan_99_Avg would be easier than remembering that cell B8 is where the average is located.

We can use the built-in naming function to let Excel assign a name to a cell. I have found this not to be great, because it doesn't usually pick the name that I would use. You still need to see how it works, so let's try it out.

Click on Cell B14 and then click on the Define Name command (see Figure 4-3).

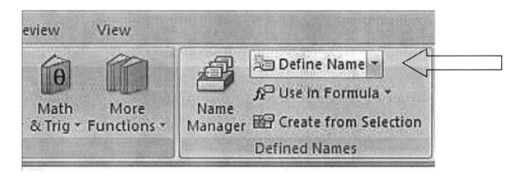

Figure 4-3

When you click on this command, the New Name dialog box comes to the screen. This is where you give the cell its name. You can also use this to assign a name to a range of cells. This is also where you tell Excel what cell, or range of cells, the name references. This can be seen in Figure 4-4.

127

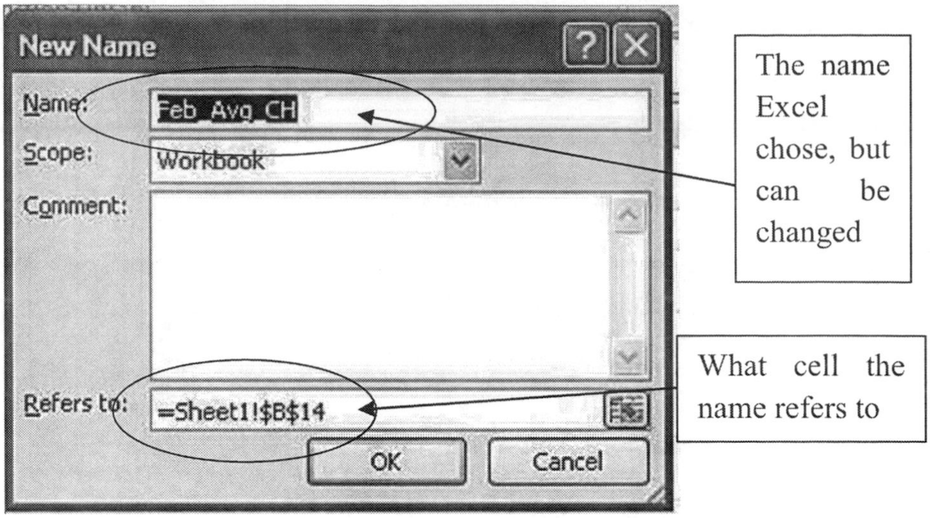

The name Excel chose, but can be changed

What cell the name refers to

Figure 4-4

You might notice that Excel chose the label of the cell next to the selected cell for the expected name. Since we are going to have eight averages for the month of February (one for each year) this is not a desirable name for the cell that is why we are going to use the name Feb_99_Avg to reference this cell.

In the Name textbox type Feb_99_Avg and then press Enter

Rename the rest of the averages for the year 1999

Save the workbook in My Documents and use the same name Church Attendance

Lesson 4 – 2   Using the Status Bar

Normally most people don't think much of the Status Bar; in fact they very seldom use it at all. This will change for most people who are using Excel 2007. There is so much information available that it would be silly not to use it. Just so you remember, the status bar is located at the bottom of the worksheet.

If necessary open Church Attendance

Right-click on the status bar

A menu will pop up showing which items are available to be displayed on the status bar. There are some items that are not checked, but you will probably want them to be. If an item is not checked, that item will not show on the status bar. The items that you will probably want to show are checked in Figure 4-5.

Customize Status Bar	
✓ Cell Mode	Ready
✓ Signatures	Off
✓ Information Management Policy	Off
✓ Permissions	Off
Caps Lock	Off
Num Lock	On
✓ Scroll Lock	Off
✓ Fixed Decimal	Off
Overtype Mode	
✓ End Mode	
Macro Recording	Not Recording
✓ Selection Mode	
✓ Page Number	
✓ Average	226.6
✓ Count	5
✓ Numerical Count	5
✓ Minimum	67
✓ Maximum	314
✓ Sum	1133
✓ View Shortcuts	
✓ Zoom	100%
✓ Zoom Slider	

Figure 4-5

If an item is not checked on your status bar menu and it is checked in the above figure, click on it to enable it.

When you are finished, click anywhere outside of the menu to close it

Now the question is: Why are we doing this?

The answer will be seen in a few moments.

Select cells B3 through B7

Look closely at the status bar (on the right side). The status bar shows how many items are selected, what the minimum value and maximum values are, what the sum of the numbers is, and what the average of the numbers is. All of this and you didn't even have to enter a formula or anything.

Figure 4-6 shows the status bar.

Average: 226.6   Count: 5   Numerical Count: 5   Min: 67   Max: 314   Sum: 1133

Figure 4-6

Lesson 4 – 3   Addition

In this lesson we are going to perform simple addition. We will learn how to reference cells in our formulas. We will also expand our understanding of the Sum function. Excel works just like you would work if you were adding numbers together (only Excel will get the right answer).

If necessary open Church Attendance

Click on cell C8

You have learned how to calculate the average attendance for a month in the last lesson. Now we are going to do it manually, so you can see what Excel is doing.

Type the following in the cell = (C3 + C4 + C5 + C6 + C7) / 5

You might need an explanation for all of the typing, so here it is.

All formulas must start with an equals sign.

Excel uses the following symbols to perform mathematical calculations:

+ for addition

- for subtraction

* for multiplication

/ for division

Excel will do multiplication and division before it does addition and subtraction, unless there are parentheses around the addition or subtraction. We need to add the weekly totals together before we divide, that is why the parentheses are around all of the additions.

After we add the weekly totals together, we want to divide by five, which is the number of weeks in January.

We could have used the actual numbers in our formula, but if we found a mistake later we would have to go back and change our formula. This way we only have to change the number in the cell and the average will update automatically.

Typing all of the cell names in the formula is a lot of work. Let's do it using an easier way.

Click on cell C14

This will be the February 2000 average.

Type the following = Sum(

Click the mouse on cell C9 and then type the colon sign :

Click on cell C13 and then type the following) / 4

When you are finished it should look like Figure 4-7.

7	Jan 05 Ch	314	172	0
8	Jan Avg CH	226.6	238.6	
9	Feb 01 CH	339	318	315
10	Feb 02 CH	301	308	306
11	Feb 03 CH	271	278	308
12	Feb 04 CH	258	313	277
13	Feb 045 CH	0	0	0
14	Feb Avg CH		= Sum(C9:C13) /4	

Figure 4-7

You will notice that Excel put a border around the group of cells that you are adding together and the cells that you clicked the mouse on are in a different color than what you typed in manually.

Press the Enter key on the keyboard

Pressing the Enter key will enter what you have typed into the cell. You will not see the formula in the cell only the result of the formula.

Save the changes and close the workbook

Lesson 4 – 4   Subtraction

Subtraction is done the same way that addition is done. Of course, you use the minus sign instead of the plus sign.

Open a new workbook

Enter the values that are shown in Figure 4-8

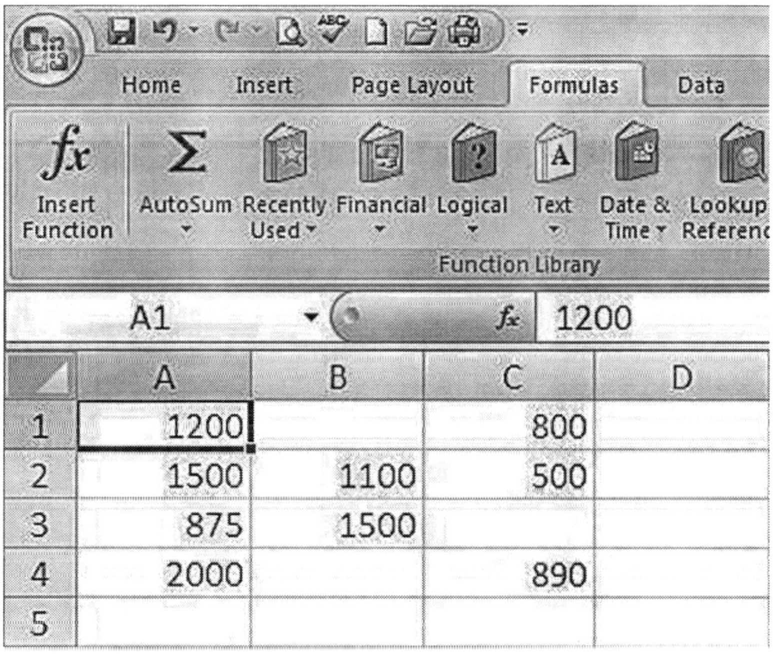

Figure 4-8

Now that the values are entered into the workbook, we can perform the subtraction that is needed.

Click on cell D1 and enter the following formula $= A1-C1$

After you enter the formula, we want to enter it, but we want cell D1 to remain the active cell. We know that if we press the Tab key, the active cell will move one cell to the right. We also know that if we press the Enter key, the active cell will move down one cell. How can we possibly enter the formula and still keep cell D1 the active cell?

Figure 4-9 will show how the screen will look after we enter the formula.

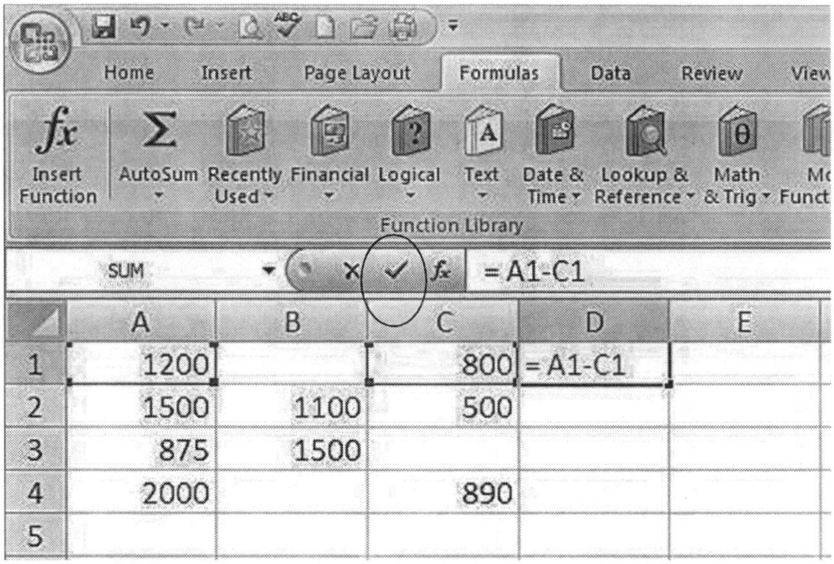

Figure 4-9

On the Formula Bar is an Enter button (circled). If you click on this button, the information will be entered into the cell and the focus will remain on the cell.

Click on the Enter Button on the Formula Bar

The cell will reflect the difference between cell A1 (1200) and C1 (800) which is 400. Notice that the formula bar will still show the formula you entered, but the cell will show the results of the formula.

Enter the formula, to show the difference between cell A2 and B2, in cell D2

The result should be 400 (1500 – 1100).

Enter the formula to subtract cell B3 from cell A3 into cell D3 and then click the Enter button

The result should be a difference that you can see quickly. The number displayed is a negative number. But what if you don't want negative numbers to be displayed like this? What if you want the more popular red number in parentheses?

Click on cell D3 (or any cell or range of cells that you would like to change the display) and click on the dialog box launcher on the Number Group of the Home Tab (see Figure 4-10)

134

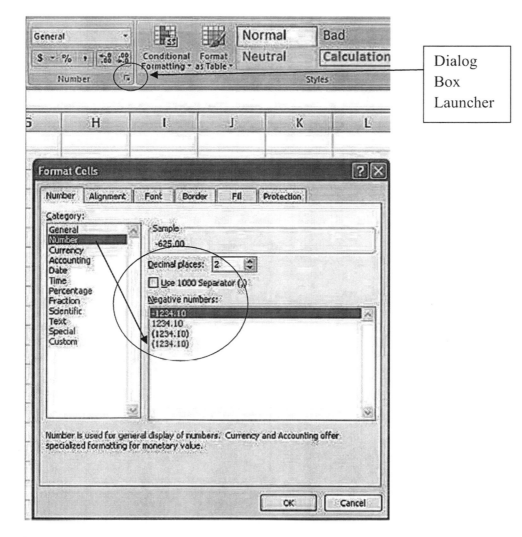

Figure 4-10

Click on the Number category and select the bottom choice for the negative numbers

This will tell Excel that you want any negative number in this cell to be in red and have parentheses around it.

Click the OK button

Cell D3 should now have a red 625 with parentheses around it.

Let's try one last thing.

Click on cell D4 and type   =

Click on cell A4

Type   —

Click on cell C4

Click the Enter button

This is just another way to do subtraction without as much typing.

Save your workbook as Subtraction

Make sure you save the file in the My Documents folder.

Lesson 4 – 5   Multiplication

Multiplication is not much different than addition or subtraction, except the operator. To multiply you use the asterisk (*) between the cell names.

In this lesson we are going to calculate how much we have earned for driving our car for the company. The company pays us thirty-five cents per mile driven.

Open a new workbook

Fill in the labels and values until it looks like Figure 4-11

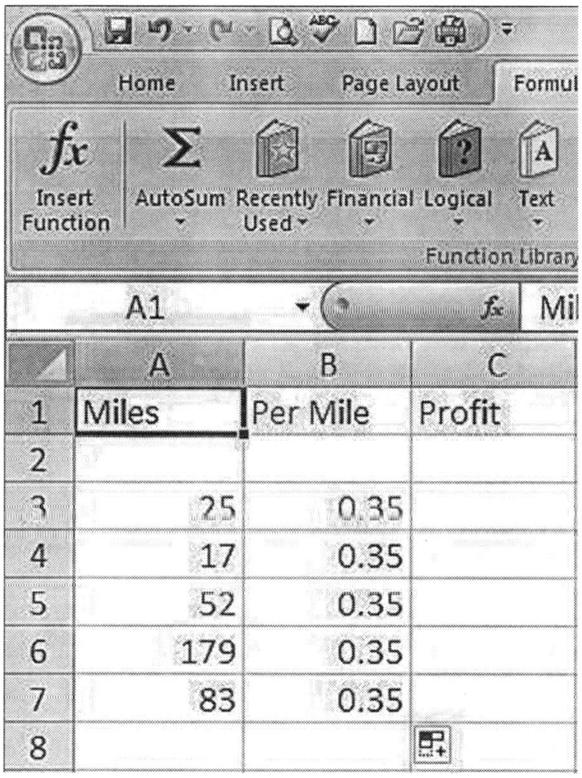

Figure 4-11

To calculate how much the company owes us, we need to multiply the number of Miles driven by the amount the company is paying us per mile.

Click on cell C3 and type the following formula = A3*B3 and then click the Enter button

The result should appear as shown in Figure 4-12.

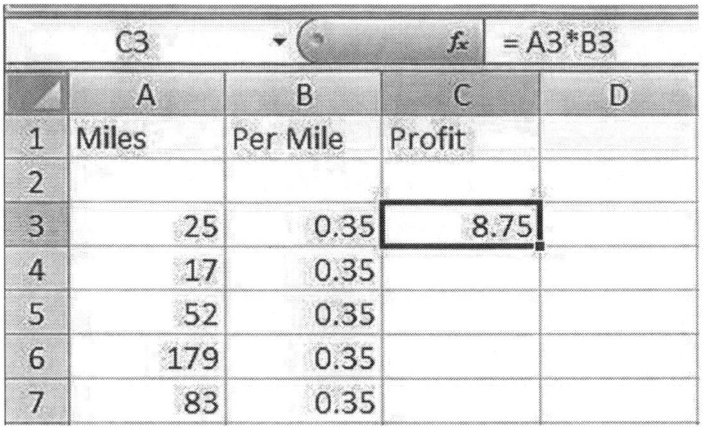

C3			f_x	= A3*B3
	A	B	C	D
1	Miles	Per Mile	Profit	
2				
3	25	0.35	8.75	
4	17	0.35		
5	52	0.35		
6	179	0.35		
7	83	0.35		

Figure 4-12

Repeat the process for cell C4

This can also be done by clicking on the cell instead of typing the cell name in the formula.

Click on cell C5 and type =

Click on cell A5 and then type *

Click on cell B5 and then click the Enter button

The result of the formula should be 18.2.

Let's try another way to find the product of two numbers.

Click on cell C6 and type the following formula = Product (A6,B6) **and then press Enter**

The result in cell C6 should be 62.65.

Calculate the amount in cell C7 using one of the above methods

Save your workbook as Multiplication

Make sure you save the file in the My Documents folder.

Lesson 4 – 6   Division

Division is nothing new either, except the symbol for division is the forward slash (/). In this lesson we will do the opposite of what we did in the last lesson.

Open a new workbook

Fill in the worksheet until it looks like Figure 4-13

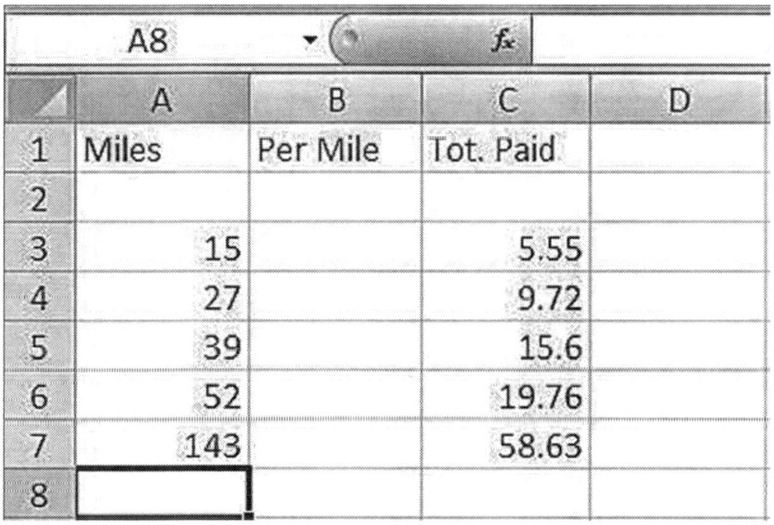

	A	B	C	D
	A8		f_x	
1	Miles	Per Mile	Tot. Paid	
2				
3	15		5.55	
4	27		9.72	
5	39		15.6	
6	52		19.76	
7	143		58.63	
8				

Figure 4-13

In this lesson, we know how much our check was and we also know how many miles we drove. Now we have to see how much per mile the company is paying us.

Click on cell B3 and enter the following formula =C3/A3 **and then click the Enter button**

As you know all formulas must start with an equals sign. The next part of the formula tells Excel to take the value in cell C3 and divide it by the value in cell A3.

Enter the formula to calculate the amount that should be in cell B4

If you entered the formula correctly, the result should be 0.36.

The correct formula should read =C4/A4.

As we have previously seen this can also be accomplished by clicking on a cell instead of typing in the cell name.

139

Click on cell B5 and type $=$

Click on cell C5 and then type the forward slash /

Click on cell A5 and then click the enter button

The resulting formula should look like this: $= C5/A5$

The result of the formula should be 0.4

Use this procedure to complete cell B6

Fill in the formula for cell B7 using one of the above methods

Save your workbook as Division

Lesson 4 – 7   How Excel Processes Math

At this time you need to be aware of how Excel processes the different mathematical calculations. Certain types of calculations are done before other types of calculations. Consider the following:

If you enter this formula into a cell = 3 + 4 *2, you might think that the resulting value shown in the cell would be 14. The reality is that the cell would show the value 11.

Excel will process the multiplication and division calculations before it processes the addition and subtraction calculations. Here is how Excel sees the formula 3 + (4 * 2). The next line, if you were writing this on paper, would look like this 3 + 8. And the result would obviously be 11.

In a like fashion, this formula would produce a value of 10; =6+12/3. Excel will see this as 6 + (12/3) or 6 + 4 which equals 10.

Excel does not have a preference between multiplication and division, so it does these calculations from left to right in the order they appear. For example, if you enter the following formula in a cell: = 6 * 4 / 8 Excel will process the formula from left to right as it is written (6 * 4) / 8 or 24 / 8 which is three. Likewise if the division appeared first it would be done first. Consider the following: =30/10*2. The resulting value would be 6. Excel will perform the division first and the multiplication second.

If you need to overwrite the standard way that Excel does the calculations, you must enclose that part of the formula in parentheses. Consider this: = (3 + 4) * 2. This will result in the value of 14 being displayed in the cell. The calculation inside the parentheses will be done before the rest of the calculations.

What is the displayed value of the following formula? = 1275 – 1225 / 25 – 2.

If you answered 1224 you would be correct. Here is the order that Excel would perform the calculations: 1225 / 25 = 49

1275 – 49 = 1226

1226 – 2 = 1224

What is the displayed value of the following formula? = (1275 – 1225) / 25 – 2.

If you answered 0, you answered correctly.

Lesson 4 – 8   Using AutoSum

Now that the four basic types of math are finished, let's move on to something that will make your life a little easier. Microsoft decided to make it easier to add numbers that are in adjoining cells by giving us a feature called AutoSum. AutoSum does what its name implies; it automatically adds values in adjoining cells. But it can do even more that just that, as you will soon see.

Open a new workbook

Fill in the cells until it looks like Figure 4-14

	A	B	C	D	E
1	200	300	400	500	
2	200		300	400	
3	300	400		500	
4	300		400	500	
5					

Figure 4-14

Click on cell A5 and then click on the AutoSum button on the Formula Tab (see Figure 4-15)

Figure 4-15

When you click the button (not the down arrow) Excel will make an assumption of what you want to add together. This may or may not be what you had in mind. If it is not, you can make changes to the selection. The result of clicking the button will look like Figure 4-16.

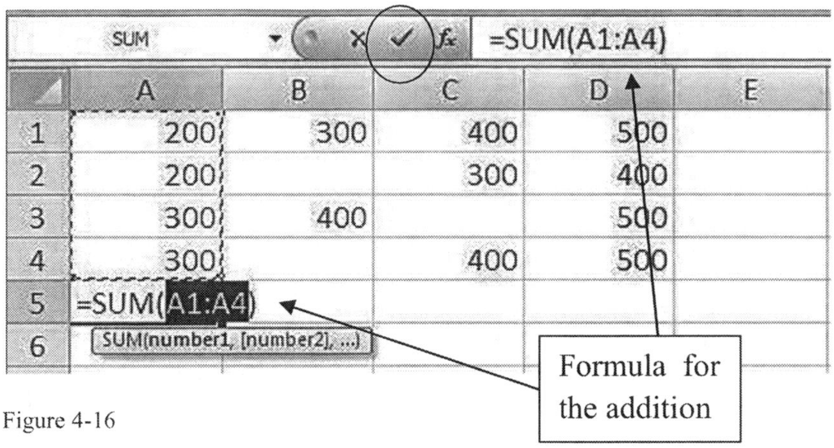

Figure 4-16

Excel put a dotted border around the cells that it thinks you want to get the sum for. You can also tell from the formula which cells are included in the addition. The (A1:A4) tells you that all of the values in cells A1, A2, A3, A4 will be added together. In this particular case, this is what we want added together. Now all we have to do is enter the formula.

Click the Enter button on the Formula Bar

Pressing the Enter key on the keyboard would also have entered the formula into the cell. The result of the addition is displayed in the cell (1000).

Click on cell B5 and perform the exact same actions and see what happens

Wow! You only came up with 400 in cell B5. What happened?

By default, Excel will skip any empty cells, and/or cells with text in them, until it comes to a cell with a value in it. It will include all cells with a value in them until it comes to another empty cell. At this point Excel will stop including more cells. If this is acceptable, all you have to do is click the Enter button. If you wish to cancel the AutoSum, press the ESC key on the keyboard.

Click on cell C5 and then click on the AutoSum button

The screen should look like Figure 4-17.

143

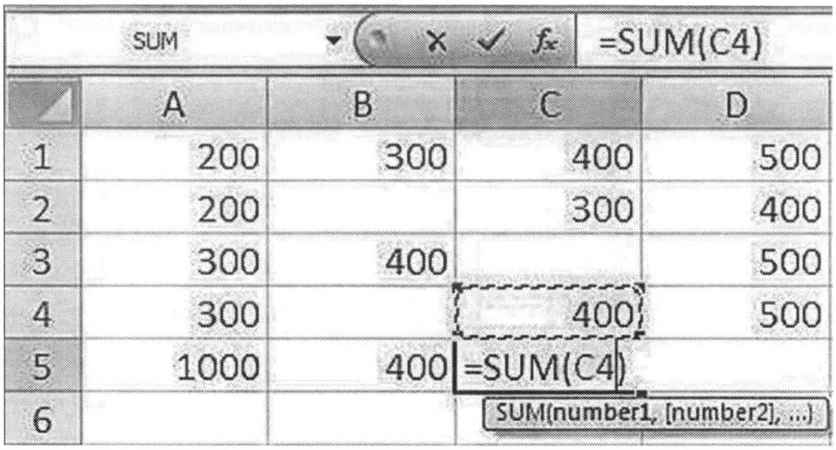

Figure 4-17

As you can see, Excel only included one cell before it came to an empty cell. If we clicked on the Enter button, the result of our addition would be 400. In this example we want to also include cell C2 in our sum. We need to drag the dotted line up to include cells C3 and C2.

Using the mouse, click inside cell C4 and drag the mouse upward until cell C2 is also inside the dotted outline

The result should look like Figure 4-18.

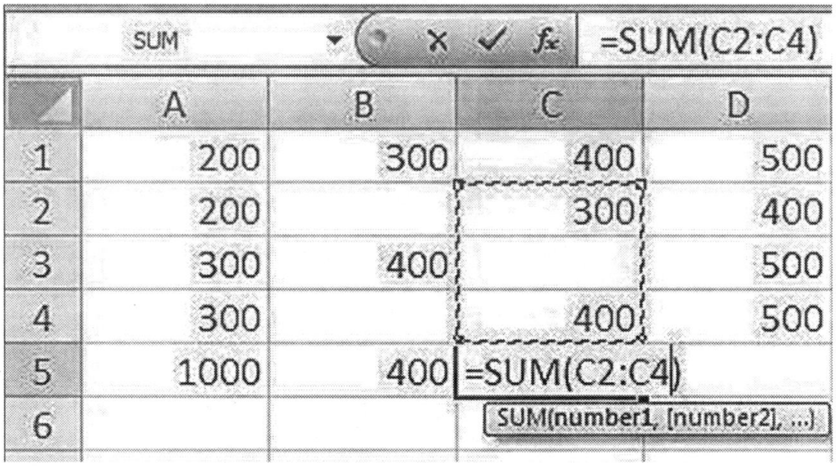

Figure 4-18

Click on the Enter button

The result, as it should be obvious, will be 700.

Let's try something a little different. As I stated earlier, the AutoSum function can do other things besides just adding numbers together.

144

Click on cell D5 and then click the down arrow on the AutoSum button

You should see something similar to Figure 4-19.

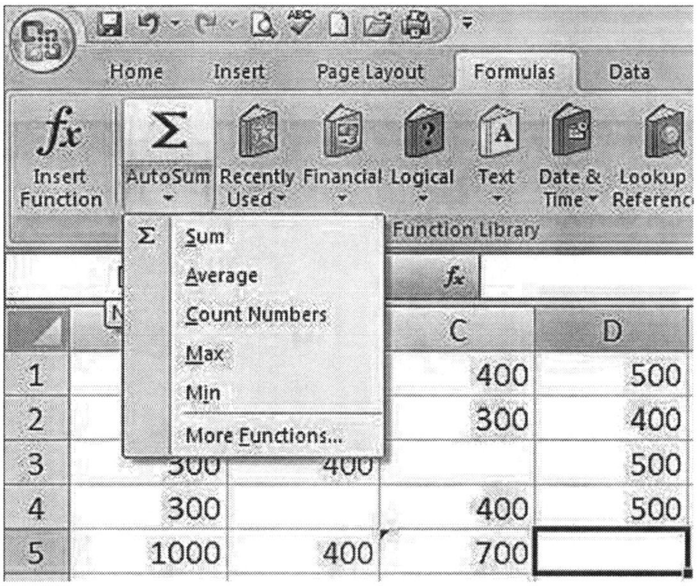

Figure 4-19

Now this to me is really cool. Not only can we get a sum of the numbers, but we can also get an average of the numbers, count the number of cells with numbers in them, get the largest number in the group, and get the smallest number in the range of cells.

Click on average and then click on the Enter button

The average of these numbers is 475. You know that to get an average, you must first add the numbers together and then divide by how many numbers there are.

Input the names Tom and Bill **in cells E1 and F1 as shown in Figure 4-20**

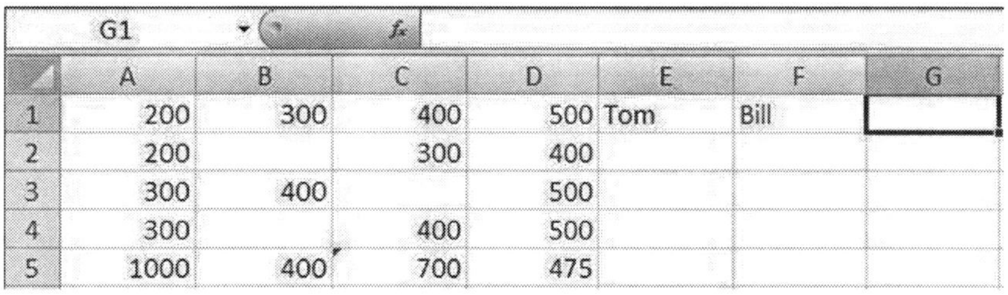

	A	B	C	D	E	F	G
1	200	300	400	500	Tom	Bill	
2	200		300	400			
3	300	400		500			
4	300		400	500			
5	1000	400	700	475			

Figure 4-20

Click on cell G1 and then click the down arrow on the AutoSum button

Click on Count Numbers and then click the Enter button

145

Why did you come up with a number 4 in the cell?

Excel counted the number of cells that have numbers in them. Cells E1 and F1 do not have numbers in them, so they were not counted. Four cells have numbers in them (A1, B1, C1, and D1).

Save your workbook as Practice_AutoSum

Make sure you save the file in the My Documents folder.

Lesson 4 – 9   Using Dates

This Lesson is only going to touch on some of the things you can do with dates. We could spend an entire chapter just on dates and using dates in our formulas. In this lesson I will try to explain how Microsoft uses dates and is able to calculate these dates in its formulas.

When you see a date, such as 7/16/2010, you see a date. When Microsoft sees the same date it see a serial number (40375). I know my first question was "Why would anyone do something as dumb as that?" As it turns out they weren't quite as dumb as I thought, in fact they were pretty smart. Let me explain.

Microsoft picked a date, January 1, 1900, in fact. This date is stored as serial number 1. Every day after that another number was added to the serial number. January 2, 1900 would be serial number 2. January 3, 1900 is serial number 3, and so on and so on. Doing this allowed Microsoft to use dates in mathematical formulas. Every date has a serial number so you can use these numbers to calculate days between dates, dates in the future, etc. The down side is that you cannot use any date before January 1, 1900 in a formula.

If I had been born on January 1, 1900, on July 16, 2010 I would be 40375 days old.

The first thing I want to show you is how to determine what any date is as far as Microsoft is concerned.

Open a new workbook

In cell A1 enter the following date 11/30/2008 and then click the Enter button

On the Home Tab and in the Number Group, click the down arrow by the Date (see Figure 4-21)

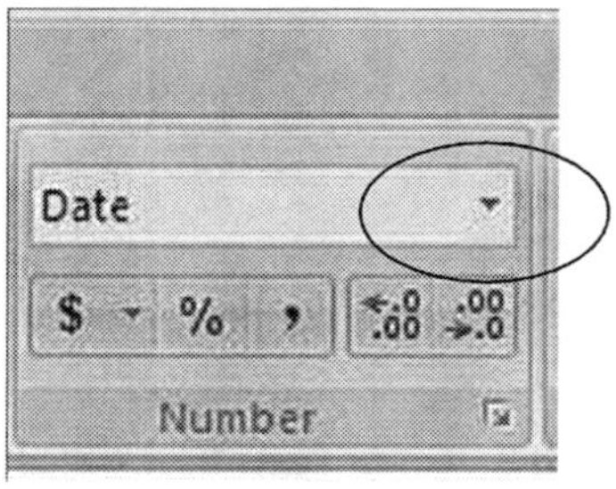

Figure 4-21

You should see a screen that looks like Figure 4-22.

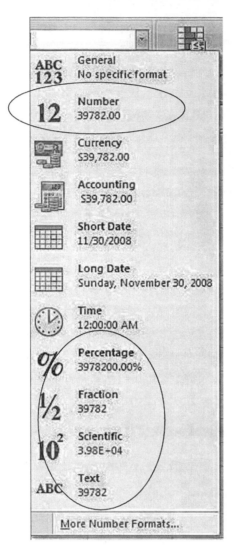

Figure 4-22

This list shows most of the ways that this date can be displayed. Notice the choices circled. The normal choice to display this as a number would be general, because there would be no decimal points in the number. What I want you to see now is you can quickly see what serial number the date refers to.

Click outside the list to make it go away

Now let's see how we can use this in a formula.

In cell B1 enter the following date 12/25/2008

Now that we have two dates, each with a unique serial number, we can calculate how many days until Christmas.

Click on cell D1 and enter the following formula =B1-A1 **and click the Enter button**

Cell D1 should now display the number of days between the two dates. There are 25 days until Christmas. You can use this type of formula to find the number of days between any two dates.

Enter today's date in one cell and your birth date in another cell

Click in a third cell and determine how many days old you are

That is a great way to determine how many days are between two dates, but what if you want to know how many <u>workdays</u> are between two dates. Let's pretend that you are working on a project and you need to know how many workdays (Monday through Friday) there are between now and the date it is to be completed. Workdays do not include Saturdays or Sundays. For this we will need a pre-made formula. This formula is called the NETWORKDAYS function.

Click on cell A3 and put in the date of 6/3/2008

This will represent the current date.

Click on cell B3 and enter the date of 12/31/2008

This will represent the day the project must be finished. Now all we need is the formula entered into another cell.

Click on cell D3 and enter this formula =NETWORKDAYS(A3,B3) **and press Enter**

There are 129 workdays between June 3, 2008 and December 31, 2008.

This works great also, except it doesn't take holidays into account. There is another formula for this. This formula is the WORKDAY formula and it requires more parameters (called arguments).

Click on cell A5 and put this date in it 3/14/2008

This will represent the current date. We are given 78 workdays to complete the project we are working on.

149

Click on cell A6 and put the number 78 **in it**

Now we need to enter all of the holidays that we do not have to work, each in its own cell.

Click on cell A7 and enter the date 3/17/2008

Click on cell A8 and enter the date 3/21/2008

Click on cell A9 and enter the date 5/26/2008

The last step is to enter the WORKDAY formula.

Click on cell A10 and enter the following formula =WORKDAY(A5,A6,A7:A9) **and then press Enter**

Excel will calculate the number of workdays (Monday through Friday), excluding all listed holidays, and give us the date we need we need to have the project finished.

We need to be finished on 7/3/2009.

If your cell displays a number such as 39997, convert it to a date

If you forgot how, refer back to Figure 4-68.

Save your workbook in the My Documents folder as Practice_Dates

Lesson 4 – 10   Using Auto Complete

Microsoft has included a feature called Auto Complete. This can be a time saver when you are typing.

Open a new workbook

In cell A1 type 3-24 **and press Enter**

You will notice that Excel changed your entry to 24-Mar.

In cell A2 type Feb 26 **and press Enter**

Excel again converts your entry to 26-Feb.

If you do not want to use this format, you can change the way it is displayed by using the drop down arrow in the Number Group of the Home Tab.

Look in the Formula bar and see what it displays (2/26/2008). The date is still there you are only displaying it differently.

There are a few things that you should be aware of when using dates in Excel.

If you do not enter a year, Excel assumes that you mean the current year.

There are a few other things you need to know if you enter a two digit year with Excel. Any date entered between 01/01/30 and 12/31/99 is assumed to be in the 20th century. Any date entered between 01/01/00 and 12/31/29 is assumed to be in the 21st century.

The Auto Complete feature can speed up data entry. This is especially true if you are entering repetitive data.

Open the Trips workbook

Click on cell A57 and type a W

Excel will realize that you are typing in the A column and see if there is a word that starts with the letter you just typed. In this case there is a William, and Excel will fill the rest of the name in the cell. If this is correct press the Enter key or click the Enter button.

There is another way to get something in a cell besides typing. You can choose from the PickList.

Right-click in cell A58 and choose Pick from the drop down list (see Figure 4-23)

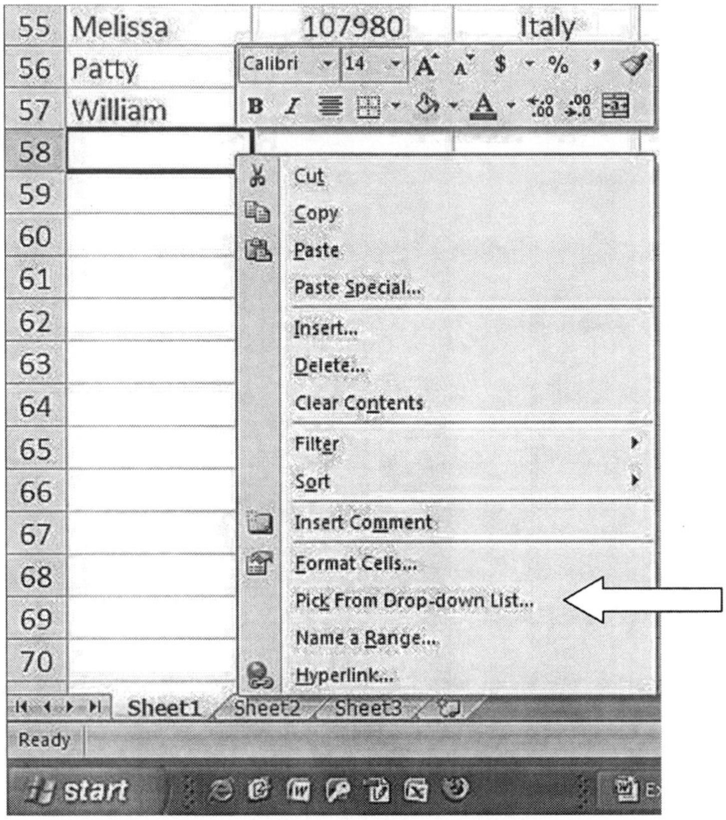

Figure 4-23

When you click on this choice there will be a list of choices from the column that you can choose from (See Figure 4-24).

53	Robert	107944
54	Thomas	107222
55	Melissa	107980
56	Patty	107517
57	William	
58		
59	Diane	
60	Mary Melissa Patty	
61	Robert Thomas	
62	William	

Figure 4-24

Click on any of the names in the list to put that name in the cell

Close the workbook without saving the changes

Lesson 4 – 11   Using AutoFill

If you are interested in something that will save a lot of time, you will be interested in the AutoFill feature. How would you like to put one formula in a cell and almost instantaneously copy it to several cells? If that sound good, you are going to like this.

Open the Church Attendance workbook

In cell C8 you need the same formula that was used in cell B8. You don't really want to retype the same formula again, do you? Excel gave us a way so that you don't have to.

Click on cell B8 and look at the bottom right corner of the cell

The bottom right corner (see Figure 4-25) has a Fill-handle on it.

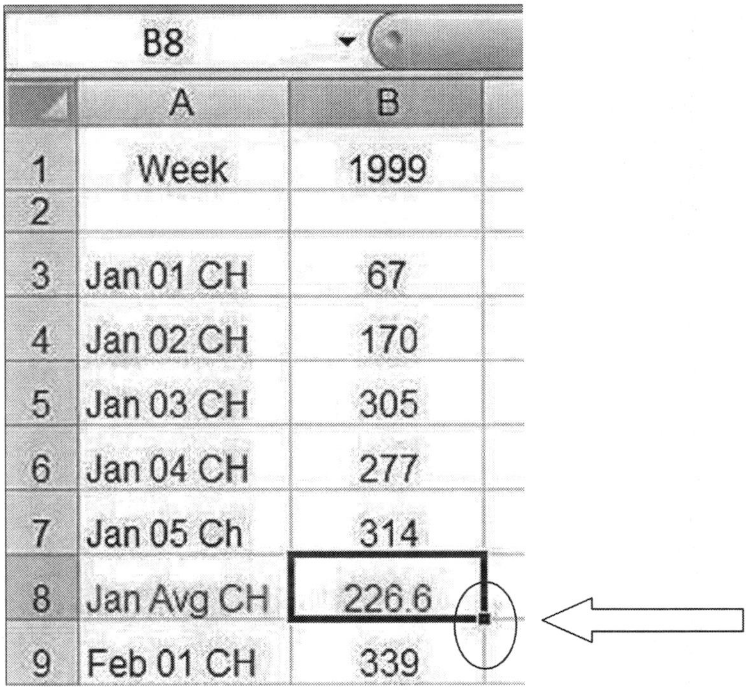

B8	▾	
	A	B
1	Week	1999
2		
3	Jan 01 CH	67
4	Jan 02 CH	170
5	Jan 03 CH	305
6	Jan 04 CH	277
7	Jan 05 Ch	314
8	Jan Avg CH	226.6
9	Feb 01 CH	339

Figure 4-25

Move your mouse over the fill handle until the pointer turns into a plus sign

Click the left mouse button and continue to hold it down as you drag the mouse over one column then release the mouse button

The formula will be copied to cell C8 and all of the cell references will also change to show cells in the C column.

Click on cell B8 and check the formula in the formula bar, now click on cell C8 and see the differences (see Figure 4-26)

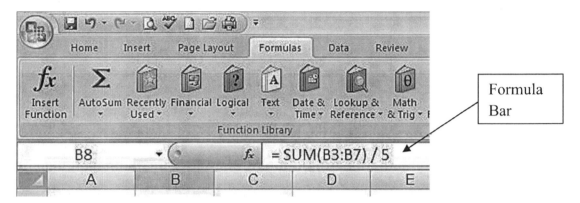

Figure 4-26

Click on cell D8 and enter this formula =SUM(D3:D6) / 4 **and then click on the Enter button**

This formula will also work in cells E8, F8, and G8.

Use the Fill-handle to copy this formula into all three cells at once

The trick is not to release the mouse button until all three cells are enclosed in the dotted line.

There is an easier way to get the same results than writing the complicated formula we used above.

Click on cell B14 and then type the equals sign =

Type AV **in the cell but do not press enter**

You will have a drop down list appear on the screen. This list will display all of the built-in commands in Excel that start with AV. This can be seen in Figure 4-27.

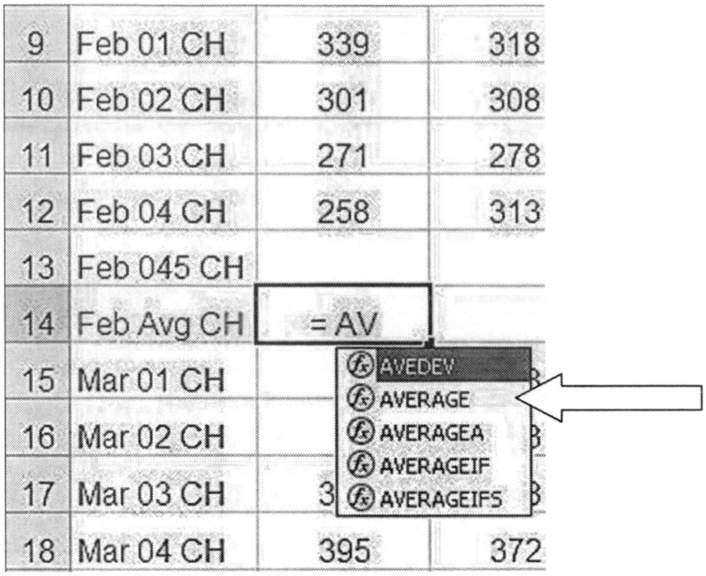

9	Feb 01 CH	339	318
10	Feb 02 CH	301	308
11	Feb 03 CH	271	278
12	Feb 04 CH	258	313
13	Feb 045 CH		
14	Feb Avg CH	= AV	
15	Mar 01 CH		
16	Mar 02 CH		
17	Mar 03 CH	3	
18	Mar 04 CH	395	372

Figure 4-27

Double-click on the word AVERAGE

The function AVERAGE is a pre-written formula which does the same thing as the formula we wrote above. It adds the numbers together and then divides by how many numbers we added together.

Inside the cell it should read = AVERAGE(

Click on cell B9 and then type the colon sign :

Click on cell B13 (even though it is empty) then type the closing parentheses)

The finished formula should look like Figure 4-28.

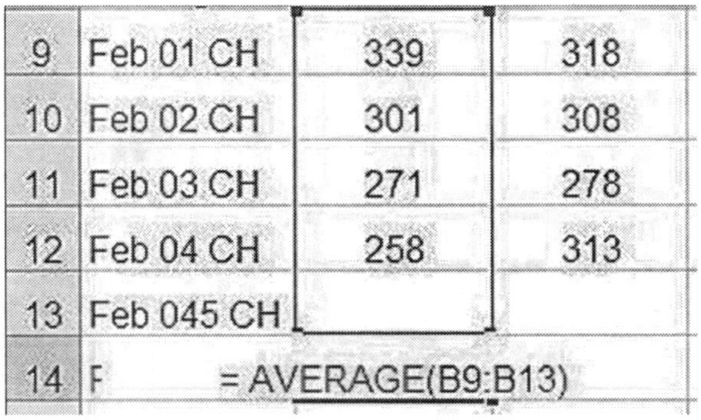

9	Feb 01 CH	339	318
10	Feb 02 CH	301	308
11	Feb 03 CH	271	278
12	Feb 04 CH	258	313
13	Feb 045 CH		
14	F	= AVERAGE(B9:B13)	

Figure 4-28

Click the Enter button

Use the Fill-handle to finish the rest of the Feb. averages then save your changes

This same procedure can be used for other things besides copying formulas.

Open a new workbook

In cell A1 type the word January **and click the Enter button**

The result should look like Figure 4-29.

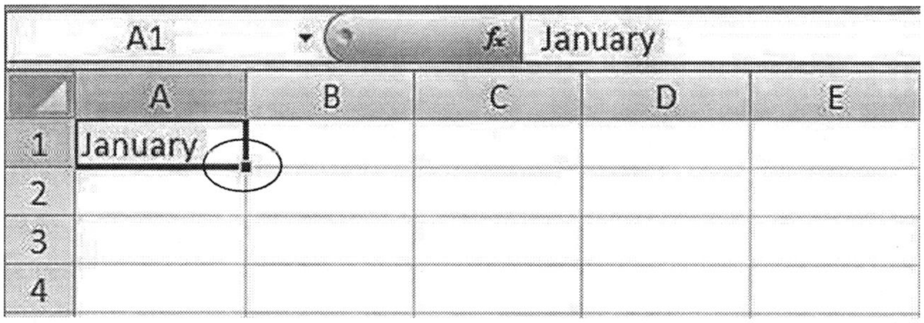

A1	▾	*fx*	January

	A	B	C	D	E
1	January				
2					
3					
4					

Figure 4-29

You need to notice that there is a Fill-handle on the bottom right corner of the cell.

Click on the Fill-handle and drag it to the right until it covers cell L1 then release the mouse button

As you drag the Fill-handle to the right the names of the months will momentarily appear as you pass through the cells. When you release the mouse button the names of the months will appear in the correct order across the top.

What if you only wanted every other month to be displayed in the cells?

Click on cell A3 and type January **and then click on the Enter button**

Click on cell B3 and type March **and then click on the Enter button**

Click on cell A3 and drag it over cell B3 (see Figure 4-30).

157

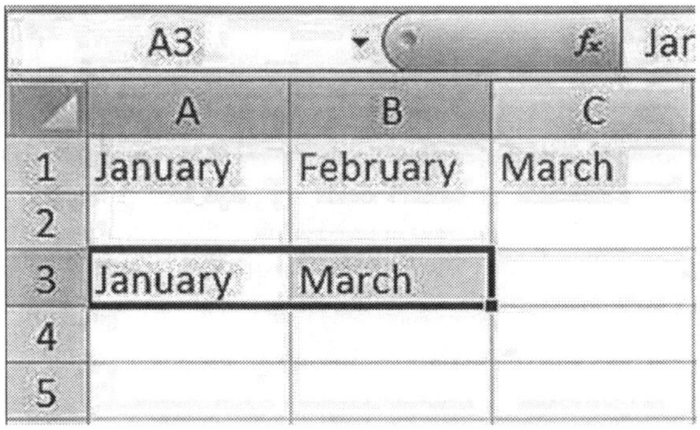

Figure 4-30

Click on the Fill-handle and drag to the right until you are in cell F3 and then release the mouse button

Now you have every other month across the cells. This works because you gave Excel a pattern to work with. Excel recognized January as the first month in a year and March as the third month. When you selected the two cells and used the Fill-handle, Excel said I will continue this pattern (odd numbers months) until you stop dragging the mouse.

Let's try one more before we quit.

Click on cell A5 and type the following number 201451

Click on cell A6 and type the following number 201452

Select both cells (A5 and A6)

Using the Fill-handle fill in the next ten cells

The result should look like Figure 4-31.

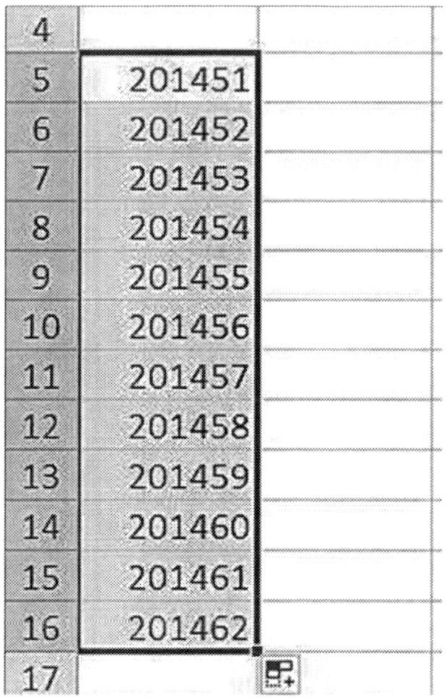

4		
5	201451	
6	201452	
7	201453	
8	201454	
9	201455	
10	201456	
11	201457	
12	201458	
13	201459	
14	201460	
15	201461	
16	201462	
17		

Figure 4-31

Close the workbook without saving the changes

Lesson 4 – 12 Cell References – Absolute / Relative

This may not seem like a lesson you would be interested in, I mean what's the big deal? Besides that, what is the difference? In this lesson we will answer both of those questions.

You already know that cells are referenced by their name. This is how Excel knows where to find a cell in a formula. You also know that if you use the Fill-handle to copy a formula, the cell references also change to match what is in the current column or row. That is because these references are relative references. If this doesn't make sense, turn back a few pages and review this again. You will find this around Figure 4-72.

An absolute reference will always reference the same cell and never change.

Let's look at a new spreadsheet to demonstrate the difference between the two types of references.

Open the workbook named Mileage Report(this file is with the downloaded files)

In this workbook we are keeping track of the mileage we charge to our customers for service calls to their business. Our charges are based on the beginning mileage and the ending mileage and charged at our per mile rate.

Click on cell D6 and subtract the beginning mileage from the ending mileage to get the total miles driven

The formula should look like this: =C6-B6

Use the Fill-handle to copy this formula to cells D7, D8, and D9

Click on cells D6, D7, D8, and D9 one at a time and check the Formula bar for any differences

Cell D6 should have =C6-B6 in the Formula bar.

Cell D7 should have =C7-B7 in the Formula bar.

Cell D8 should have =C8-B8 in the Formula bar.

Cell D9 should have =C9-B9 in the Formula bar.

In plain English here is what Excel said to itself:

Take the value that is in the cell to my left and subtract the value that is in the cell that is two places to my left. Any time we copy it to another cell, it will say the same thing. Take the value that is in the cell to my left and subtract the value that is in the cell that is two places to my left.

That is why we end up with a different value in each cell. The cell reference is relative to where the formula is located.

Are you ready for the next step?

Click on cell E6 and enter this formula =D6*E1 **and click the enter button**

Use the Fill-handle to copy the formula into cells E7, E8, and E9 and then click the Enter button

What happened? It didn't work properly. Why?

The answer is the same as why the last copy did work. Here is what Excel said to itself:

Take the value that is in the cell to my left and multiply it by the value that is in the cell located five cells above me. That sounds good, except what is in the cell that is five cells above cell E7? That's right nothing. There is nothing there to multiply by. What about five cells above E9, there is text in that cell not a value. We can't multiply by text.

What are we going to do about this? The answer is simple; we need to always reference the same cell even if the formula is moved. We need an absolute reference.

Click on cell E6

Click on the formula shown in the Formula Bar

Make sure the insertion point (the vertical flashing line) is on the reference to cell E1 (see Figure 4-32).

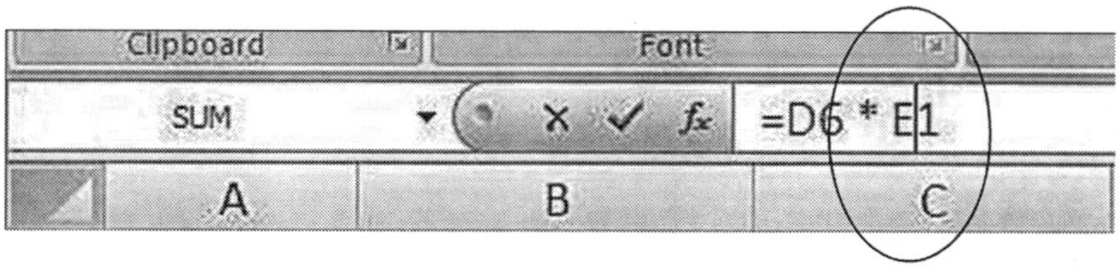

Figure 4-32

Press the F4 key on the keyboard

161

The formula will change to read =D6*E1. The dollar signs tell Excel to always use this cell, no matter where the formula is moved to or copied to. This is an absolute reference.

Click the Enter button

For the moment nothing seems to change. This is because you copied the formula before it had an absolute reference.

Use the Fill-handle to copy the formula to cells E7, E8, and E9

Now all of the cells should have a value in them.

Note: you could have manually typed the dollar signs into the formula the first time you typed it in the cell, instead of going back and doing it after the fact.

Save the Mileage Report spreadsheet in the My Documents folder

Lesson 4 – 13   Formula Auditing

If you are ever having trouble getting a formula to work, this lesson might help you. In this lesson you will learn how to show all of the formulas in your worksheet and how to have Excel evaluate the formulas for you. You will also see how to show which cells are involved in a formula.

Open the Mileage Report2 workbook that is in the downloaded files

This worksheet has one cell that does not have a value in it. We need to find out why there is no value in this cell. The first thing we will want to do is show all of the formulas in the worksheet.

Click on the Formulas Tab and find the Formula Auditing Group (see Figure 4-33)

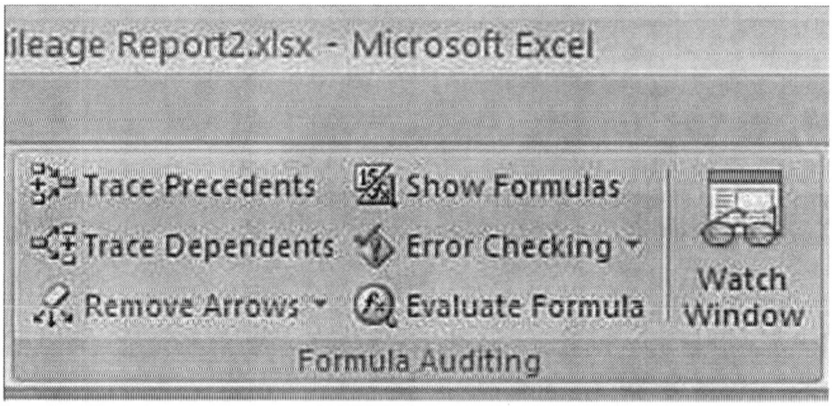

Figure 4-33

In the Formula Auditing Group click on Show Formulas

The screen should change and now show all of the formulas for all of the cells, not the results from the formulas as we normally see.

Now that we can see the formulas, it should be obvious why there is no value in cell E9. Let's pretend that we don't see the obvious for a few moments, or you will miss all the cool stuff you can do from here.

Click on cell E9

Your screen should have a section that looks like Figure 4-34.

D	E
Per Mile Charge	2.9
Total Miles	Amount Charged
=C6-B6	=D6*E1
=C7-B7	=D7*E1
=C8-B8	=D8*E1
=C9-B9	=D10*E1

Figure 4-34

From here we can see a couple of things. The first thing is that the first cell referenced in the formula is highlighted in blue and the second cell referenced is highlighted in green. The actual cells are bordered in blue and green. It should be obvious that the first cell referenced is empty and therefore our results are not what we expected.

Just for fun, let's pretend that we didn't notice this. We can have Excel evaluate our formula for us.

Click on Evaluate Formula

A new dialog box will come onto the screen. This dialog box will allow us to walk through the entire formula and see what is happening. Figure 4-35 shows a snapshot of the Evaluate Formula Dialog Box.

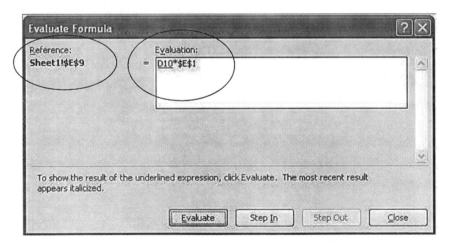

Figure 4-35

If you click the Evaluate button, Excel will check to see if the reference is valid. It is a valid cell so that won't help us. What we want to do is see what is in the cell and if it is valid for the formula.

Click the Step In button

From this view of the dialog box, we can see what is in the cell referenced in the formula.

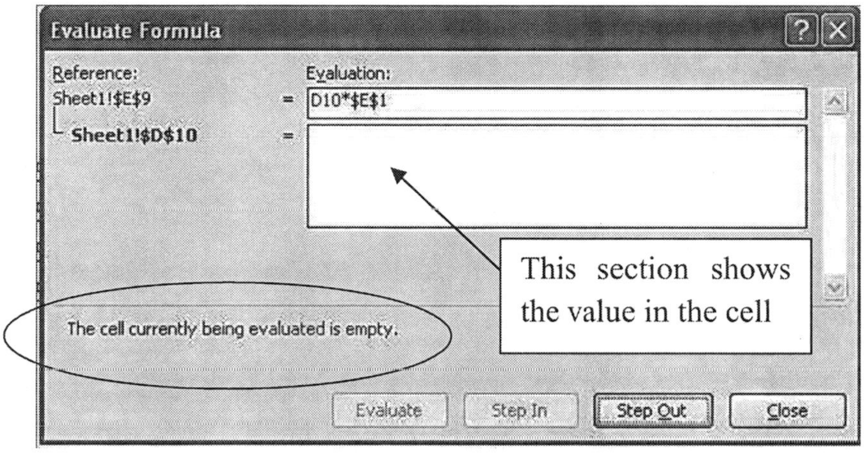

Figure 4-36

Now we know for sure why we have no value in cell E9, our formula references a cell that is empty.

Click the Step Out button to continue with the evaluation

Click the Step In button to see what is in the next cell referenced in the formula

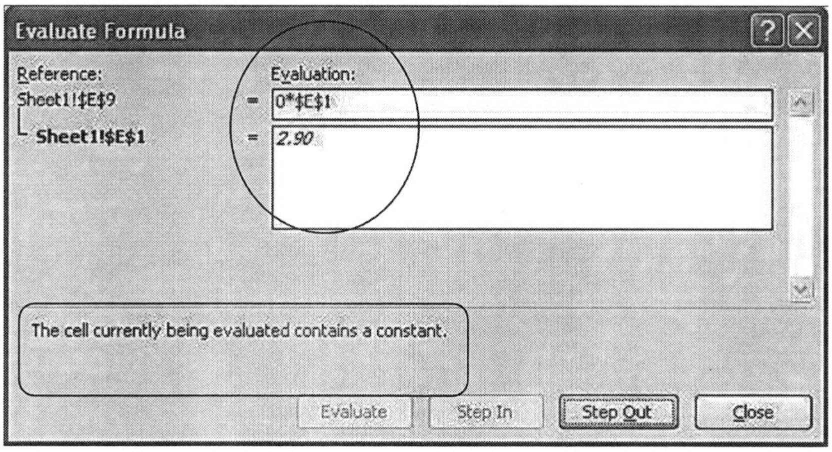

Figure 4-37

Click Step Out to see the last referenced cell

165

Click Evaluate to see the results of the formula

Click on the Show Formula button a second time to go back to the original screen

How would like to see something else that is pretty nifty?

Click on cell D7

Click on the Trace Precedents button in the Formula Auditing Group

This will put arrows on the screen to show what cells affect the value of the currently selected cell.

Click the Remove Arrows button

Click on the Trace Dependents button

This will put arrows on the screen to show what cells are affected by the value of the currently selected cell.

If we had done this at the very beginning, we would have seen that the reason cell E9 was not filled in with a value was because one of the cells referenced was empty.

Click the Remove Arrows button

Save the workbook in My Documents as Mileage2

Close the workbook

Lesson 4 – 14   Using the If Function

This lesson will teach you how to put some conditions on your formulas. What if you worked as a store manager and you promised your salespersons a 1% bonus at the end of the month, but only if they sold a certain amount of merchandise? You can let Excel do the calculations for you.

Open the Sales Bonus workbook that is part of the downloaded files

A few things to get you up to speed on this workbook; all salespersons receive the same base salary, each salesperson will receive a bonus only if the reach their quota. If they reach their quota they will receive their base salary plus a 1% bonus.

Now let's make Excel do it for us.

The If function contains three arguments

Logical Test
The value returned if the test is true
The value returned if the test is false

As usual, there is an easy way and a hard way to get this done: let's do the hard way first.

Click on cell D3 and enter the following formula =IF(C3>=B3, C3*0.01, 0)

Now the explanation:

Here is the first step

> If the value in cell C3 is greater than or equal to the value in cell B3

> That is if the sales are greater than or equal to the sales quota.

Second step

> Then the value in cell D3 is equal to the value in cell C3 times .01

> That is the bonus amount is 1% of the sales.

Third step

Or else the value in cell D3 is 0

That is if the first step is not true the bonus is zero.

Only one of the last two steps can happen. If the first step is true, then the second step is performed. If the first step is false, then the third step is performed.

This is also known as an If, Then, Else statement. If the condition set forth in the first step of the formula is true, then do the second step, or else do the third step.

Click on the Enter button

Since the sales were only 149,000 and the quota was 150,000 there is no bonus to put into cell D3.

Using the Fill Handle copy this formula to cell D4

There is a bonus figured for cell D4 because the sales were higher than the quota. The bonus is 1% of 130,000 dollars or $1,300.00.

Now let's try the easier way, well not a lot easier but a little easier.

Click on cell D5

Type the Equals sign followed by the letter i

Double click on the word IF **in the drop down list**

We still have to fill in the parts of the formula. We are still going to be looking for the same circumstances as we had in the last one. What we are going to be looking for is to see if the sales were greater than or equal to the quota. That means that we still have to make the comparison.

Click on cell C5

Using the keyboard, enter the following >= **and then click on cell B5**

What this is telling Excel to do is to compare the two values and see if the value in C5 is equal to or greater than the value in cell B5.

Using the keyboard enter a comma

168

This next part tells Excel what to do if the test made in the first step is true. If the value in cell C5 is equal to or greater than the value in cell B5 then put this in cell D5.

Using the keyboard enter C5*0.01

This tell Excel that if the test was true and the value of the cell C5 was equal to or greater than the value of cell B5 then perform the math specified and put this value in cell D5.

Using the keyboard enter a comma

The next part of the formula tells Excel what to do if the test was not true and the value in cell C5 is NOT equal to or greater than the value in cell B5.

Using the keyboard enter a zero

This tells Excel that if the value of C5 is not equal to or greater than the value in cell B5 put a zero in cell D5.

Press the Enter key

The sales were higher than the quota, so there is a bonus figured in this cell.

Using the Fill Handle copy this formula to the remaining cells

I hope you can see that by putting conditions on the formula, you can start using some of the true power of Excel.

Save this file in the My Documents folder

169

The default cell names can be changed. This can be done manually in the Name Box or by using the Define Name command on the Formulas Tab. The name can reflect a single cell or a group of cells.

The Status Bar can show the user valuable information. You can choose what is shown by right-clicking on the Status Bar and choosing the options you want displayed.

All formulas must start with an equals sign.

To add the values of two cells you list the cell names with a plus sign between them.

To subtract the values of two cells you list the cell names with a minus sign between them.

To multiply the values of two cells you list the cell names with an asterisk between them.

To divide the values of two cells you list the cell names with a forward slash between them.

Excel processes multiplication and division before addition and subtraction. Multiplication and division are considered equal to Excel and they are processed in the order they appear from left to right. The same is true for addition and subtraction. To deviate from this you must enclose the mathematical calculation you want done first in parentheses.

AutoSum will automatically add numerical values in a row or column until it finds a cell without a numerical value in it.

Dates are recorded as serial numbers in Excel. Because of this they can be used in calculations.

AutoComplete will let you easily complete entries in a cell.

AutoFill will allow you to drag formulas and numerical sequences to other cells.

Relative cell references will change depending on where the formula is copied or moved to.

Absolute references will always reference the same cell.

Formula Auditing will let you show the formulas in the cell instead of the results of the formula. You can also see which cells are affected by the selected cell and which cells affect the selected cell.

The IF function will allow you to put conditions on a formula.

Chapter Four   Quiz

1) Mar98 is a name that you can give to a cell. **True or False**
2) Excel will not allow spaces in a cell name. **True or False**
3) Write the formula for adding cells B2, B3, and B4 together.
4) Auto Subtract is a function on the Formulas Tab. **True or False**
5) The Asterisk * is the sign for division in Excel. **True or False**
6) What is the result of entering this formula: 275+125/5-200?
7) AutoSum will stop adding if it comes to an empty cell. **True or False**
8) If you wanted to find the number of workdays (Monday through Friday) between two dates, would you use the NETWORKDAYS formula or the WORKDAY formula?
9) What is the name given to the bottom right corner of a cell that you would use to drag a formula to another cell?
10) Pressing F4 on the keyboard while the mouse has been clicked inside a cell reference in the Formula Bar will make this an absolute reference. **True or False**

Chapter Five   Managing Files

If you are one of those people who have your entire monitor screen filled with icons and your documents folder is so full you have trouble finding the correct file, this chapter may change your computer life. Consider this, would you have a filing cabinet stuffed full of papers and none of them in file folders? I highly doubt if you would do that. So why would you do that to your computer?

Think of your computer as a large filing cabinet. If you had all of your documents separated and tucked neatly away in folders keeping like documents together, your frustration might just go down one level.

In this chapter we will work on organizing your files into folders.

Lesson 5 – 1   Making New Folders

Before you can organize your files, you need a place to keep them. On the preceding page I stated that your computer is like a filing cabinet. In your filing cabinet at home, you probably keep your papers and files in a file folder. I would imagine that you keep papers that are similar in the same folder. I doubt that every piece of paper has its own folder. I would suggest that you use this same line of thought for storing files on your computer.

Open Windows Explorer

Windows Explorer can be found by clicking on Start, and then All Programs, and then Accessories, then Windows Explorer (see Figure 5-1).

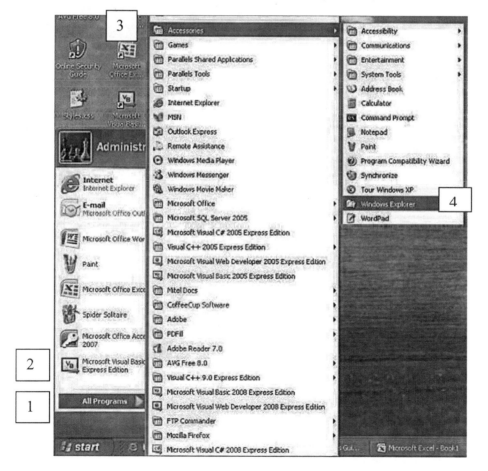

Figure 5-1

When you open the Windows Explorer program the first thing you should see is the My Documents folder and its contents. This is shown in Figure 5-2.

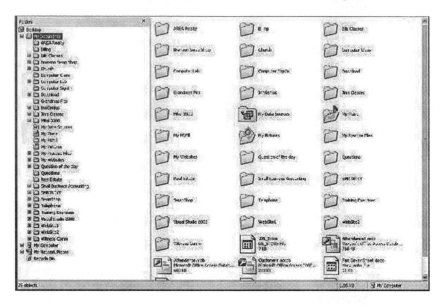

Figure 5-2

Your screen will not look like the one shown in the above figure. This is a snapshot of the screen on my computer, yours will look different. What I want you to see is that there a several folders under My Documents and not a lot of individual files.

Figure 5-3 shows a closer look at the left side of the screen.

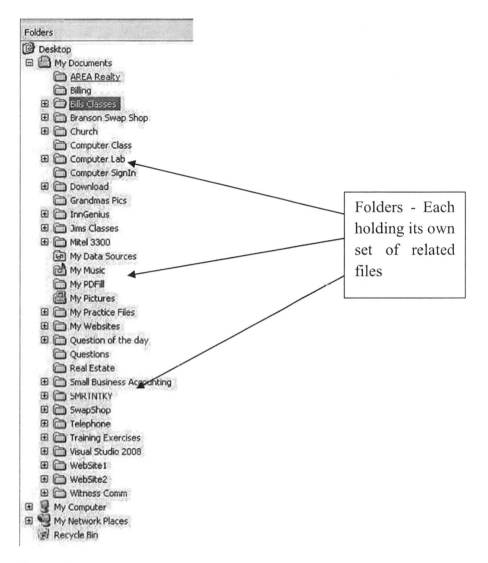

Folders - Each holding its own set of related files

Figure 5-3

Inside each folder are related files. The name of the folder gives me a clue as to which files I have put inside the folder. As an example, the folder that has "Grandmas Pics" on it has pictures relating to my mother.

For this lesson we are going to make a folder to hold all of the workbooks you have been using.

Make sure the My Documents folder is highlighted and click on File then New, then New Folder

This is shown in Figure 5-4.

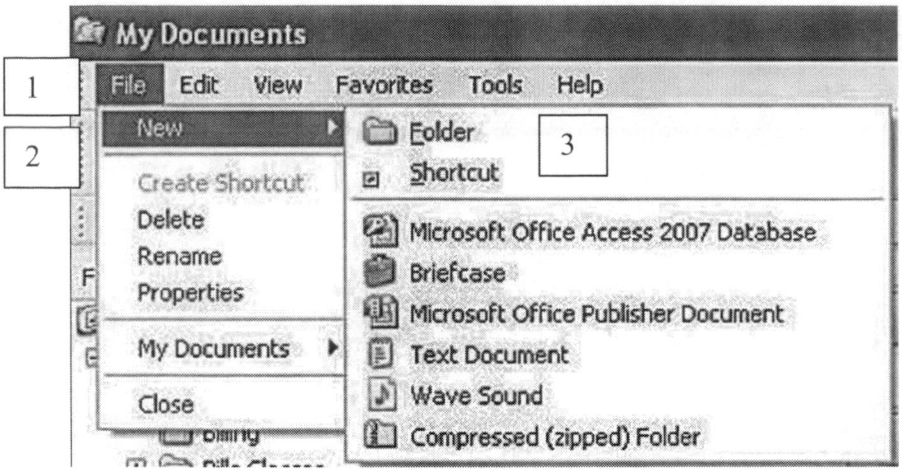

When you click on the New Folder option from the menu, you will see that a new folder appears under the My Documents folder (see Figure 5-5).

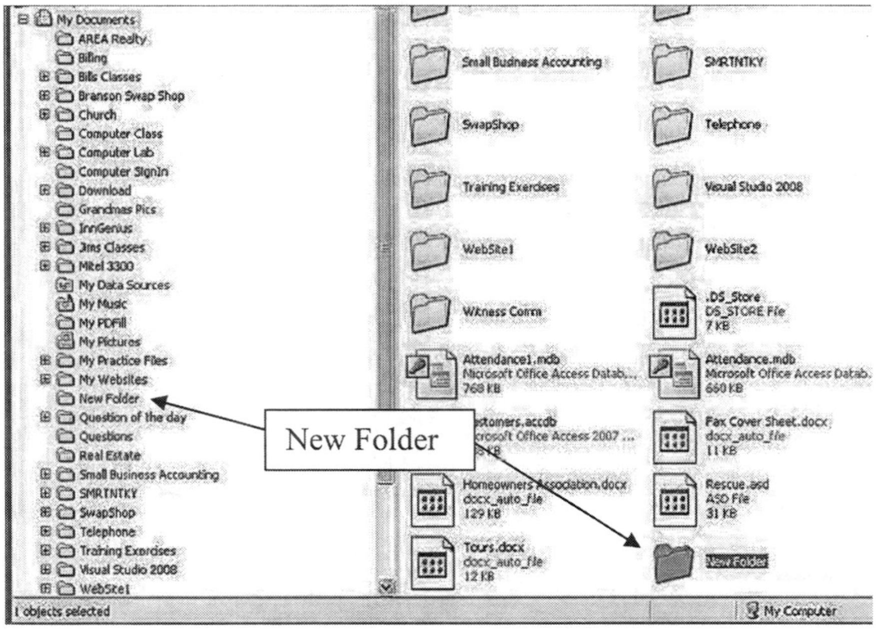

Figure 5-5

The name of the New Folder is highlighted in blue (New Folder) and is ready for you to start typing a new name for the folder. You should give the folder a new name, something that will describe what is kept inside of it.

Using the Keyboard type Excel Class **and then press Enter**

The new folder is now named Excel Class. You will also notice that it may be in a new location. Windows can automatically arrange the files into alphabetical order. To see what is inside of a folder you need to double-click on it.

Using the mouse double-click on the Excel Class folder

The folder will open and show you its contents. Wow! There is nothing in there. The folder is empty, and now you are wondering what good is an empty folder. Just like at home, we need to put something in the folder for it to be of any use.

In the next lesson we will put something inside this folder. For this lesson this is all you need to know.

You can create a new folder anytime you want by using this method. You also don't have to create the folder under My documents. You can single-click on any folder and then create a new folder inside of it.

<h1>Lesson 5 – 2   Moving Files</h1>

Let's put the workbooks we have been using inside the folder. In this lesson I will show you how to move a file (workbook) from one folder to another folder.

I asked you way back in lesson 1-10 to make sure you saved all of your workbooks in the My Documents folder. Now we are going to move them to the Excel Class folder. If you didn't save them in the My Documents folder, I don't know where they will be, you are going to have to look for them. To find them you may have to click on Start and then search and then all files and folders, then you can enter the name of the workbook (such as Trips) and then click search. In the results section, you will be able to see where the file is located.

The rest of this lesson is going to assume that your workbooks are located in the My Documents folder.

Click on the My Documents folder

You should see, somewhere in there, a set of files that look something like the ones shown in Figure 5-6.

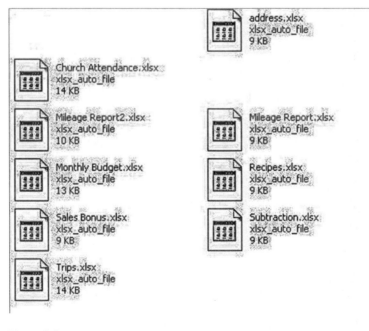

Figure 5-6

Click on the file named Trips and drag it over to the folder named Excel Class and then release the mouse button

As you pass over the folders, each one will highlight to let you know that if you release the mouse button this is the folder your file will be moved to. When the folder named Excel Class is highlighted, you can release the left mouse button. The file named Trips should now be gone from the My Documents folder. Let's see if it in the Excel Class Folder.

Double-click on the Excel Class folder to open it

The results should look like Figure 5-7.

Figure 5-7

Using this same procedure move the remaining workbooks to the Excel Class folder

The workbooks are named: Address, Church Attendance, Division, Mileage Report, Mileage Report2, Monthly Budget, Multiplication, Recipes, Sales Bonus, and Subtraction.

That is all there is to moving files from one folder to another.

Just for the sake of having something to do, we want to keep our workbooks in the Excel Class folder, but we also want the trips file to be in the My Documents folder. To accomplish this we will need to copy the file from one folder to another, not move it.

Double-click on the Excel Class folder to open it if it is not already open

Your screen should look similar to the one shown in Figure 5-8.

Address	C:\Documents and Settings\Administrator\My Documents\Excel Class		Go

Folders

- Desktop
- My Documents
  - AREA Realty
  - Billing
  - Bills Classes
  - Branson Swap Shop
  - Church
  - Computer Class
  - Computer Lab
  - Computer SignIn
  - Download
  - Excel Class
  - Grandmas Pics
  - InnGenius
  - Jims Classes
  - Mkel 3300

Trips.xlsx
xlsx_auto_file
14 KB

address.xlsx
xlsx_auto_file
9 KB

Church Attendance.xlsx
xlsx_auto_file
14 KB

Mileage Report2.xlsx
xlsx_auto_file
10 KB

Mileage Report.xlsx
xlsx_auto_file
9 KB

Monthly Budget.xlsx
xlsx_auto_file
13 KB

Recipes.xlsx
xlsx_auto_file
9 KB

Sales Bonus.xlsx
xlsx_auto_file
9 KB

Subtraction.xlsx
xlsx_auto_file
9 KB

Figure 5-8

Right-click on the file Trips and select copy form the menu (see Figure 5-9)

Trips.xlsx
xlsx_auto_file

Scan with AVG Free

Open With ▶

Send To ▶

Cut

Copy

Create Shortcut
Delete
Rename

Properties

Figure 5-9

Right-click on the folder My Documents and choose Paste from the menu (see Figure 5-10)

Figure 5-10

Now the file workbook (Trips) is located in both folders (Excel Class and My Documents). That is all there is to copying a file.

Copy all of the files that you downloaded from the website to the Excel Class folder.

Chapter Five   Review

Organizing your files and folders can make your life easier. Putting similar files together inside of a folder can make the files easier to find.

Clicking the File command on the Menu Bar will allow you to create a new folder. You should give the new folder a new name and the name should give you a clue as to its contents.

You can drag files from one folder to another folder. You can also right-click on a file and choose cut or copy from the shortcut menu.

Chapter Five   Quiz

Create a new folder inside the My Document folder.

Name the folder My Files.

Move two of the Excel files to the new folder.

Chapter Six   Formatting

In this chapter I will try to give you the ability to present a more attractive and easier to read spreadsheet.

In this chapter we will learn how to adjust the height of the rows and the width of the columns. We will change the size and color of the text. We will also learn how to align the text inside the cell, and add borders and patterns to a cell.

Lesson 6 – 1   Adjusting Row Height and Column Width

Before we start changing what is seen inside the cells, let's cover something that will also affect how we see what is inside the cells. The two things that will affect how we view what is inside the cell are the height of the rows and the width of the columns. If the row is not tall enough, the numbers and text will look cramped and scrunched up. If the column is not wide enough, the text and numbers will not fit inside the cell.

Adjusting the height and width is very simple, as you will see.

Open a new blank workbook

Move the mouse pointer over the number 1 on the left side of the workbook

When the mouse pointer is over the number it will appear as an arrow.

Slowly move the mouse down until it gets to the bottom of the row and turns into a black plus sign

When the mouse pointer turns into the plus sign, you can adjust the height of the row.

Click and hold the left mouse button as you drag the line at the bottom of the row down

As you drag the mouse down, a screen tip will be displayed to show you the new size of the height. This number will change as you move the mouse. The dotted line will show you where the new row height will move to. See Figure 6-1 for an example of this.

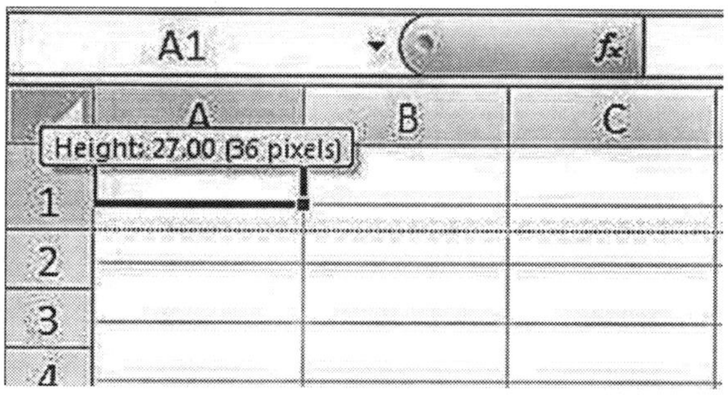

Figure 6-1

When the height gets to 27.00, release the left mouse button

185

The row height is now set to 36 pixels.

Adjusting the column width is just as easy.

Move the mouse pointer over the letter A at the top of the first column

When the pointer is over the letter it will show as a downward pointing arrow. When it gets to the column line it will change into a black plus sign as it did with the row line.

Move the pointer to the right until it changes into the plus sign

When the mouse pointer changes into the plus sigh, the width of the column can be adjusted.

Click and hold the left mouse button down and drag the mouse to the right

As you drag the mouse to the right, a screen tip will be displayed to show you the new width of the column. This number will change as you move the mouse. The dotted line will show you where the new width will be moved to. See Figure 6-2 for an example of this.

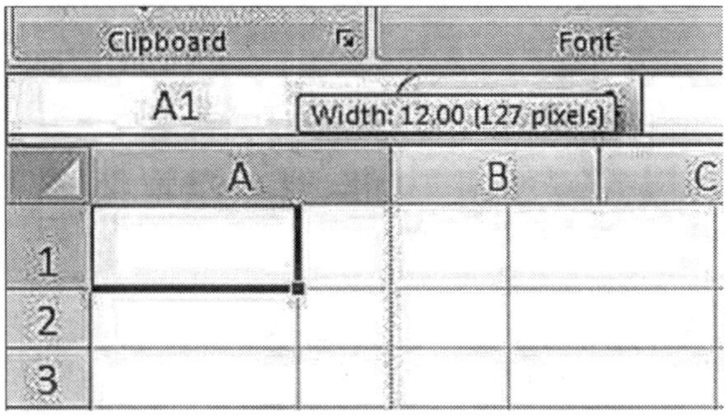

Figure 6-2

When the width reaches 12.00 release the left mouse button

The column width is now set.

Note: You can have Excel automatically adjust the size of the column for you. If there is text in the cell and you want to adjust the width of the column to automatically adjust to the correct width to display the text, move the mouse pointer just as we discussed above and when it changes to the plus sign double-click the mouse. You can also refer back to chapter three to use the Ribbon to automatically adjust the column width.

Close the workbook without saving your changes

Lesson 6 – 2 Formatting Text

Create a new workbook called Income & Expenses

In this workbook we will use some of the same figures we used in the Monthly Budget workbook. This time we will also include our income. Fill in your worksheet until it looks like the worksheet in Figure 6-3.

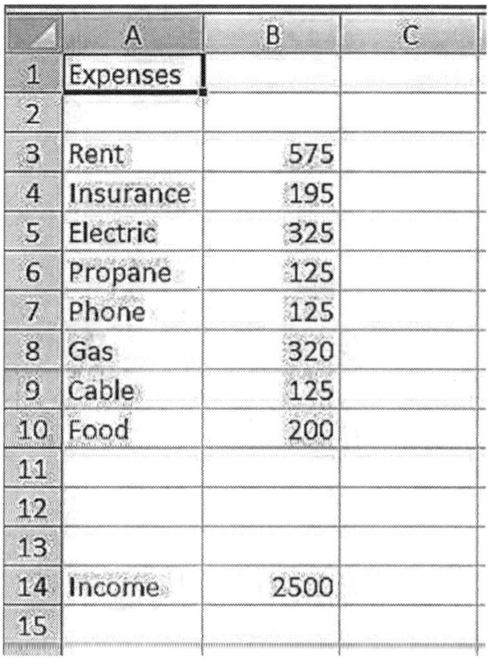

	A	B	C
1	Expenses		
2			
3	Rent	575	
4	Insurance	195	
5	Electric	325	
6	Propane	125	
7	Phone	125	
8	Gas	320	
9	Cable	125	
10	Food	200	
11			
12			
13			
14	Income	2500	
15			

Figure 6-3

This is going to be our basic spreadsheet. Now let's dress it up a little. First we should change the font size and color of the Expense and Income labels.

Click on cell A1 and change the font to Stencil (See Figure 6-4)

Note: you can also type the name of the desired font in the Font Name box instead of using the drop down menu.

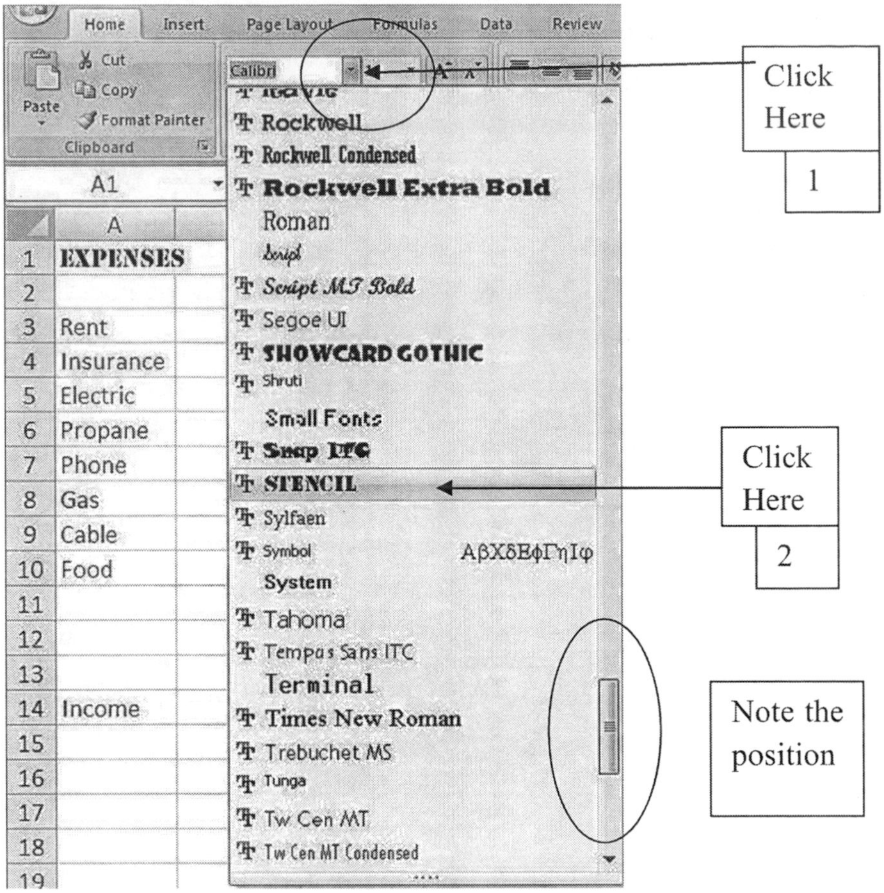

Figure 6-4

Repeat this process for cell A14

These two cells represent the two main labels and therefore need to be in a larger font that will make them stand out from the rest of the spreadsheet.

Click on cell A1 again

With the cell selected, we can change the size of the font.

Click on the down arrow on the font size command (see Figure 6-5)

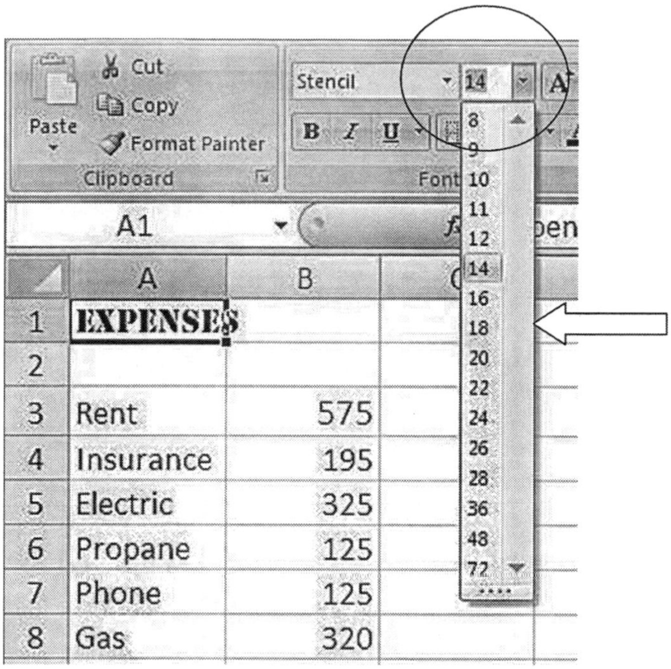

Figure 6-5

Click on the number 18 to make the font size larger

Repeat the above process for cell A14

The first thing that you should notice is that the label INCOME does not fit into the cell. The label Expenses also does not fit into cell A1, but there is nothing in the cell next to it (B1) so the text overflows and is visible. That is the difference. Cell B14 has a value in it, so the text cannot overflow. To fix this we need to make the column width larger.

Using the procedure we used in Lesson 6-1, make the column width equal to 12.00

The labels will now fit inside the cells.

Now let's move the labels and values in the Expenses over one column to the right.

Select the cell range A3 through B10 (see figure 6-6).

189

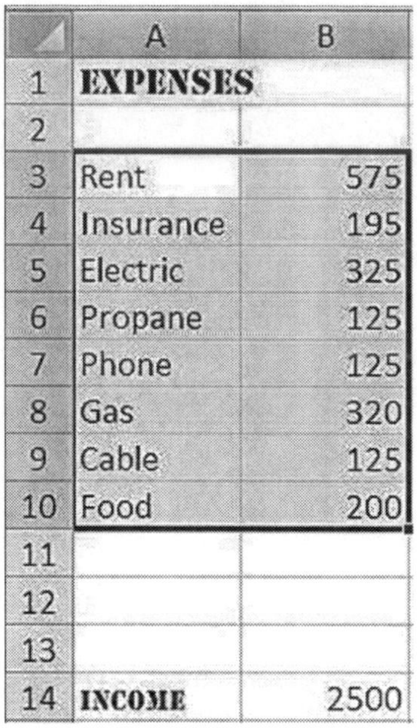

	A	B
1	**EXPENSES**	
2		
3	Rent	575
4	Insurance	195
5	Electric	325
6	Propane	125
7	Phone	125
8	Gas	320
9	Cable	125
10	Food	200
11		
12		
13		
14	**INCOME**	2500

Figure 6-6

As you know, this can be done by clicking (and holding the left mouse button down) on cell A3 and dragging the mouse over to cell B10 and then releasing the left mouse button.

Click the Cut button

This button, you will remember, is on the Home Tab and in the Clipboard Group.

Click on cell B3 and then click the Paste button

This button, you will remember, is also on the Home Tab and in the Clipboard Group. The result should look like Figure 6-7.

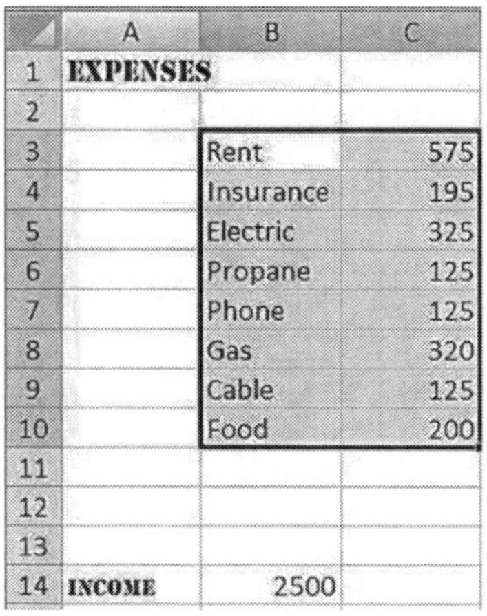

	A	B	C
1	**EXPENSES**		
2			
3		Rent	575
4		Insurance	195
5		Electric	325
6		Propane	125
7		Phone	125
8		Gas	320
9		Cable	125
10		Food	200
11			
12			
13			
14	**INCOME**	2500	

Figure 6-7

190

This will separate the expenses from the labels. Now let's make some adjustments to the labels for the actual expenses.

Select cells B3 through B10 and then click the Italic button (see Figure 6-8)

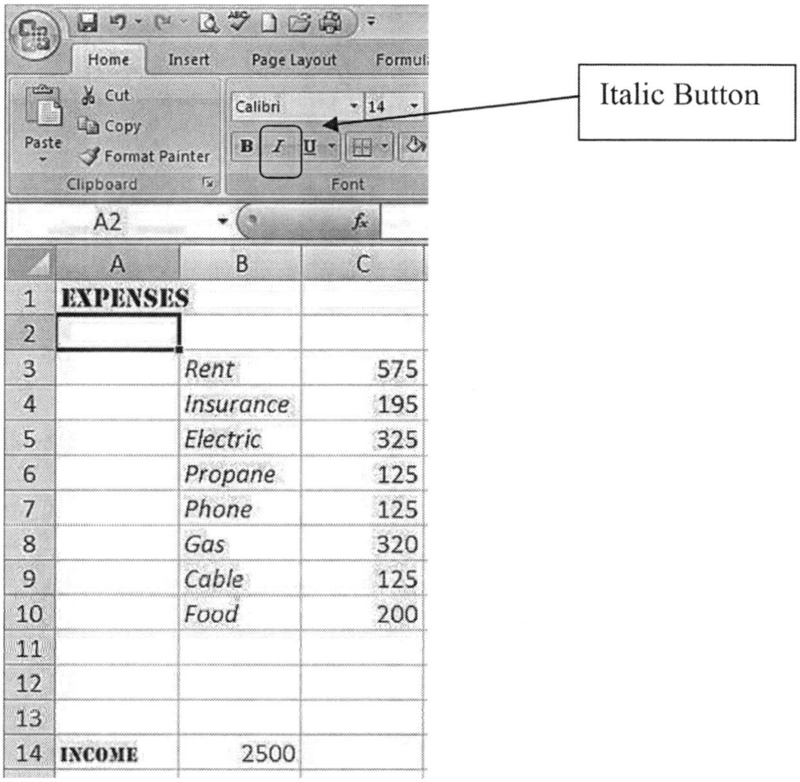

Figure 6-8

The result is that the text is now slanted and again it stands out from the rest of the spreadsheet. Since they are an outgoing expense, let's make the text red.

With the cells still selected, click on the Font color down arrow (see Figure 6-9)

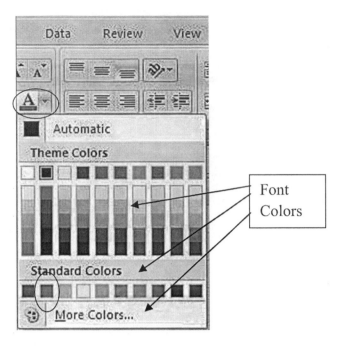

Figure 6-9

Click on the Red choice under Standard Colors

For the most part, this is what you will do when formatting text. There are other things and we will address these in the upcoming lessons in the rest of the chapter.

At this time we should look at the other options available in the Font Group of the Home Tab.

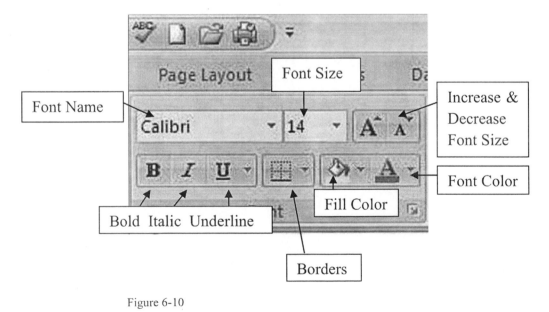

Figure 6-10

We have discussed changing the Font to a different style. The name of the selected font is shown in the Font Name text box.

The Font Size determines how large the selected font will appear. Fonts are measured in points and each point is 1/72 of an inch. The larger the number is, the larger the font is.

Directly to the right of this are two buttons, one increases the size of the font and one decreases the size of the font. Every time you click the button, the font size will increase or decrease by one size.

On the bottom part of the Font Group you can change the font from regular to bold, italic, and/or underlined. You can also add borders to the cell, change the background color of the cell, and change the color of the font.

Save your changes to the workbook

Lesson 6 – 3   Formatting Values

When formatting values, we can do the same things with numbers that we can do with text. We can change the size of the font, we can change the type of font used, we can make it bold, italic, or underlined, and we can add borders, just as you can with text.

There are also some things that we can change just because it is a number.

If necessary open the income & Expenses workbook

Figure 6-11 shows you the Numbers Group of the Home Tab.

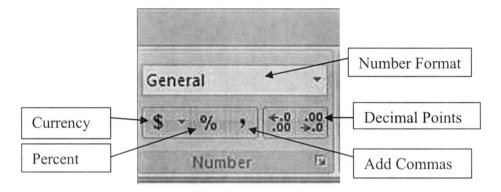

Figure 6-11

We have discussed the number format in a previous lesson. I am referring to chapter Four when we discussed dates. The options are still the same as they were in Chapter 4. By clicking on the down arrow you will be able to see the different ways that the contents of the cell can be displayed.

The Currency ($) button can be used to display the number as a currency. The down arrow will allow you to use the currency from other countries instead of the American dollar.

The percentage (%) button will allow you to display the number as a percentage.

The comma button will display the number with a comma separator between the thousands and it will add two decimal points to the display.

The Decimal points command will allow you to add a decimal point or remove a decimal point from the number. The right half will remove a decimal point and the left half will add a decimal point.

Click on cell C3

Click the dollar sign part of the currency button

The way the number is displayed is changed to represent currency. The dollar sign is present and there are two decimal points shown (See Figure 6-12).

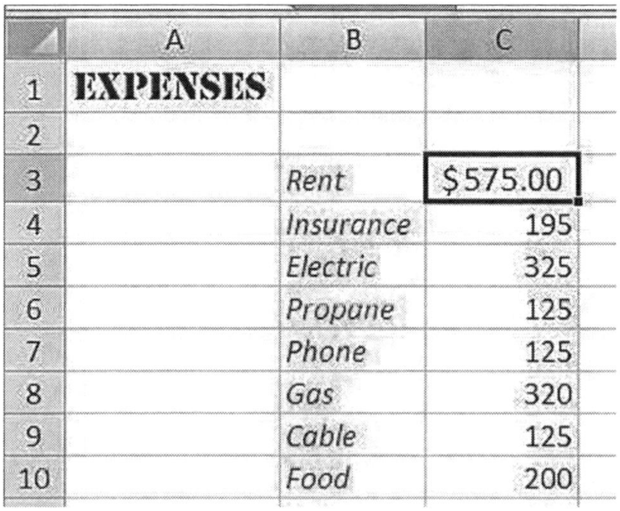

	A	B	C
1	**EXPENSES**		
2			
3		Rent	$575.00
4		Insurance	195
5		Electric	325
6		Propane	125
7		Phone	125
8		Gas	320
9		Cable	125
10		Food	200

Figure 6-12

Repeat this for cell C4

We will change the others a little later using different techniques in the next few lessons.

Save your changes

Just for fun, let's see how the other buttons in the Number Group affect the way the values are displayed.

Click on cell C5

Click the percent button

Wow! I bet that is not what you expected to get is it?

Figure 6-13 shows the result of clicking the % button.

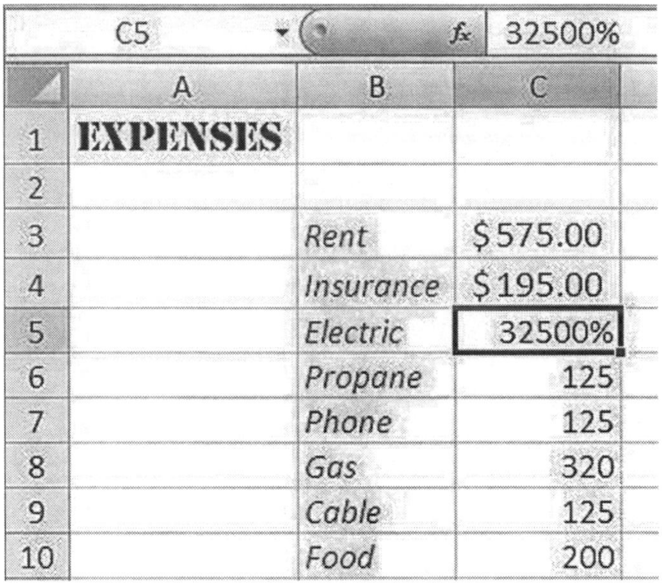

Figure 6-13

Okay it is time for an explanation, right?

If you had the number 0.10 in a cell and clicked the percentage button, Excel would display this number as 10%, which is what it is. If you had the number 1 in a cell and clicked the percent button, Excel would display the number as 100%, which is what it is. So here is the rule; If there is a number in the cell, Excel will multiply it by 100 to get the percentage. Therefore you may not get the results you expect when you click on the % button.

On the other hand, if the cell is empty and you select it and then click on the % button, numbers equal to and larger than 1 are automatically displayed as percentages. A number that is smaller than 1 is multiplied by 100 to get the percentage. An example might explain this better:

Click on cell D5 and click the % button before you enter a value

Enter 10 in the cell and press Enter

The result shows 10% in the cell. The number 10 is larger than 1, so the result is shown as 10%.

Change the value to .1 and Press Enter

The result also shows 10% in the cell. The number .1 is smaller than 1 so it is multiplied by 100 before it is shown as a percent.

Note: The moral of the story is to be careful when you click on the % button.

196

Click on cell C6 and then click on the Increase decimal button (see Figure 6-14)

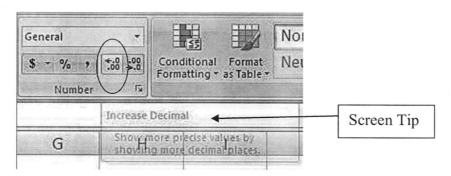

Figure 6-14

The result is that the number will change from 125 to 125.0. Clicking it a second time will result in the number being 125.00.

Click on the Decrease decimal button until the display shows 125 again

Close the workbook without saving the changes

Lesson 6 – 4   Using the Dialog Box

If necessary open the Income & Expenses workbook

We have looked at how to format text and numbers from the Home Tab. Now we will look at how to make these changes using the dialog box. Figure 6-15 shows where the Dialog Box Launcher is located.

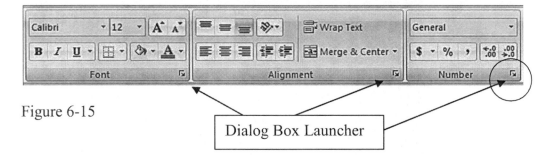

Figure 6-15

Dialog Box Launcher

Normally I would go through these one at a time, only they all launch the same dialog box.

Click on the Number Dialog Box Launcher button

The Format Cells Dialog Box will jump onto the screen as shown in Figure 6-16.

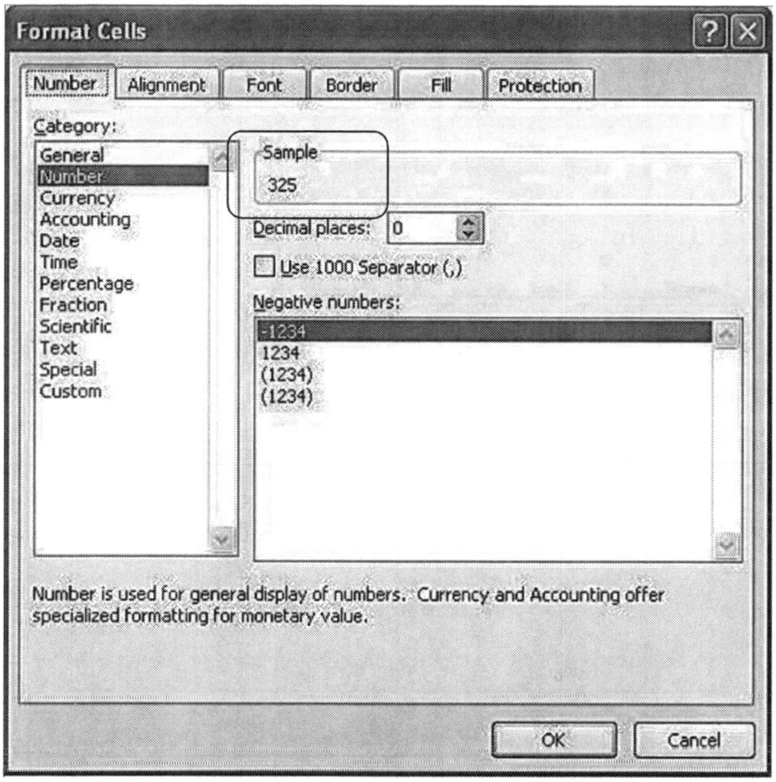

Figure 6-16

It doesn't matter which dialog box launcher you click on, the same Format Cells Dialog Box will come to the screen. The only difference is which tab is shown on the front.

Click on each of the choices under Category and watch the Sample Area to see the changes

Excel has several built-in number types as you can see. For this part of the lesson we will concentrate on the last two types listed. We will be looking at the special and custom choices.

Click the Cancel button to close the dialog box

Click on cell E5 and enter the following number 2125551212 **and then click the Enter button**

Click on the Number Dialog Box Launcher

Click Special then Phone Number then OK

The number displayed in the cell will change to show a phone number with parentheses

199

(212) 555-1212.

Click on cell E7 and enter 471275 and then click the Enter button

Excel tries to format the entry before it displays it in the cell. If Excel recognizes it as a date, or a time, or a percentage, or something familiar like that, it will automatically format it. Since it is not something Excel recognizes it will be displayed as it was typed.

Let's pretend that you decide that all employee ids need to be in the following format 47-1275.

There is no predefined format for this; we will need to make our own custom format.

Click on the Number Dialog Box Launcher and select Custom

The Dialog box is shown in Figure 6-17.

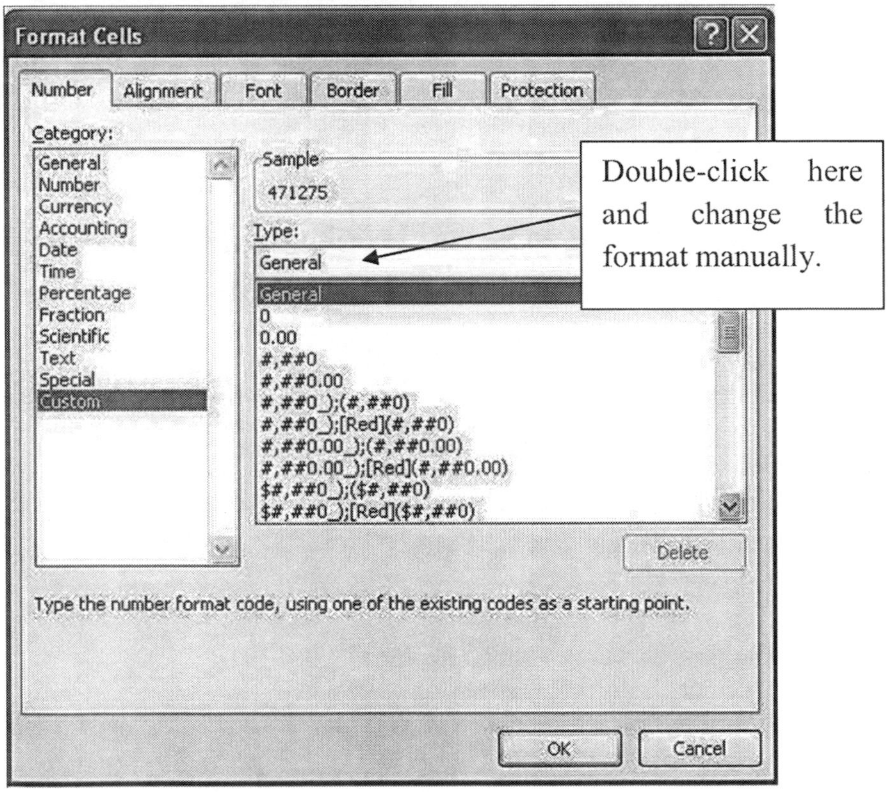

Figure 6-17

Double-click on the word General

The word General will be highlighted in black and this is where you will type the format you want to use.

Type the following ##-#### and then click the OK button

The number will change from 471275 to 47-1275.

There are so many possibilities of ways to format numbers that I don't know where to start. The # symbol is a placeholder as well as the 0 being a placeholder. The main difference is weather the extra zeros will be displayed if the entered number has fewer digits that the format calls for.

If you entered the number 8.9 but wanted it displayed as 8.90 you would enter the following in the custom section #.00

If you entered 8.9 and the custom format was #.## the number would be displayed as 8.9.

Using the zeros instead of the # sign will cause the extra zero to be displayed if fewer digits are entered than required.

You could add comas to the format such as: #,###.##

If you would like more information on custom numbers, click on the help button and type custom numbers in the search section. There are several pages of examples of how to write custom formulas. The Help button is shown in Figure 6-18.

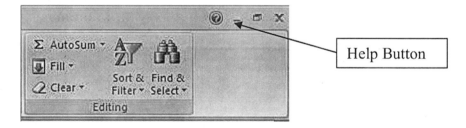

Figure 6-18

The following Figure shows the other most commonly used tab in the Format Cells Dialog Box. This would be the Font Tab.

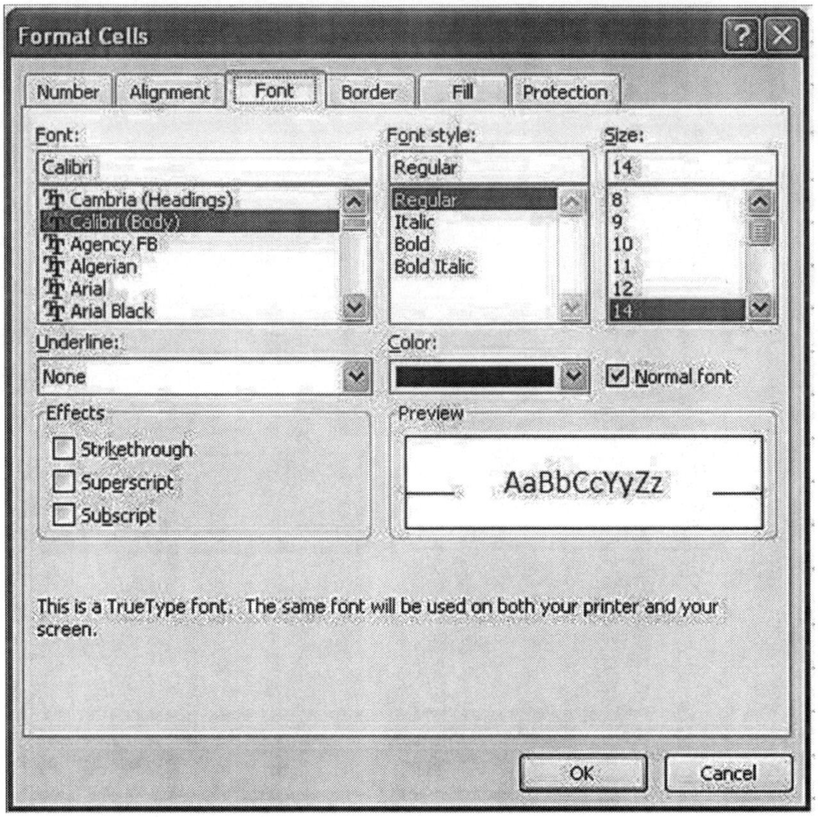

Figure 6-19

These choices are also available from the Font Group on the Home Tab. The other dialog boxes will be shown with future lessons.

Lesson 6 – 5   Using Cell Alignment

Aligning the text and numbers inside of the cells is something you will do with almost every workbook. We therefore need to go over how to use this part of the Ribbon.

Figure 6-20 Shows the Alignment Group of the Home Tab of the Ribbon. We will be referring to this during the remainder of the lesson.

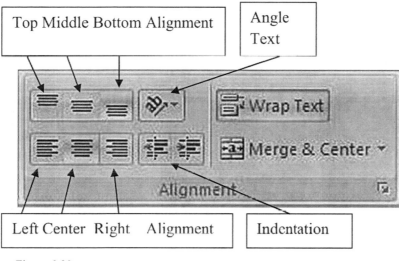

Figure 6-20

Open the Income & Expenses workbook if necessary

I want you to notice that by default text is aligned to the left side of the cell and numbered are aligned to the right side of the cell. Let's change the default on some of the cells.

Select cells B3 through B10

We are going to change the alignment from the left side to the right side.

Click on the Right Alignment button

The text should appear as shown in Figure 6-21.

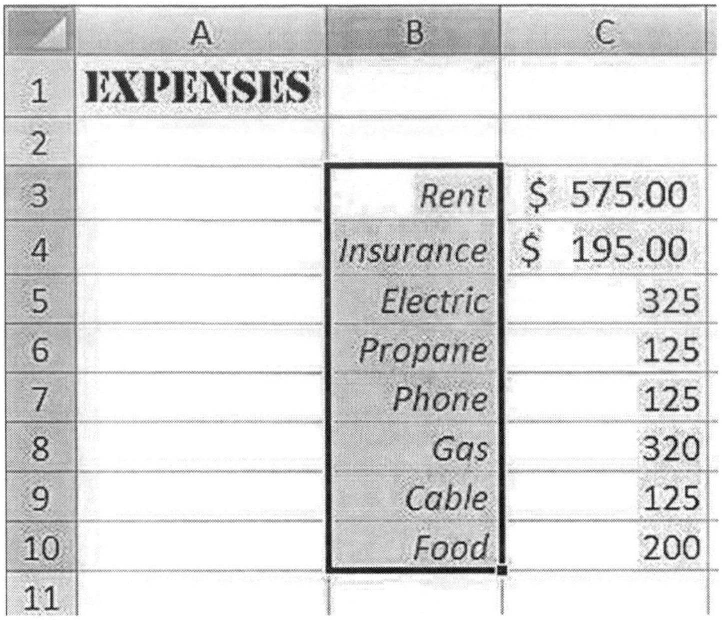

	A	B	C
1	**EXPENSES**		
2			
3		Rent	$ 575.00
4		Insurance	$ 195.00
5		Electric	325
6		Propane	125
7		Phone	125
8		Gas	320
9		Cable	125
10		Food	200
11			

Figure 6-21

Click on Column A and then Center align the text

All of the text in column A will be in the center of the cells. It may not be visible in cell A1, but it is visible in cell A14.

This same alignment can be done for the cells with numbers.

Click on cell C5 and center the number

The text should look like this:

Figure 6-22

Increase the size of row 1 to 42.75 as we did in Lesson 6-1

With the row larger, we will be able to see the difference that the other three alignment buttons make on information typed into cells.

Click on cell A1 and then click on the top, center, and bottom align buttons

You will be able to see the differences between each of the three buttons and how it aligns the text vertically in the cell.

There are times when you will want one cell to cover several columns, but only in one row. You might want a heading at the top of the worksheet, but only for that one row.

Select cells A1, B1, C1, and D1

When all four cells are selected, you can merge the cells together to form one cell. This is a pretty good way to have one cell be large enough for a heading and leave the rest of the cells at their normal size.

Click the Merge and Center button

The result should look like Figure 6-23.

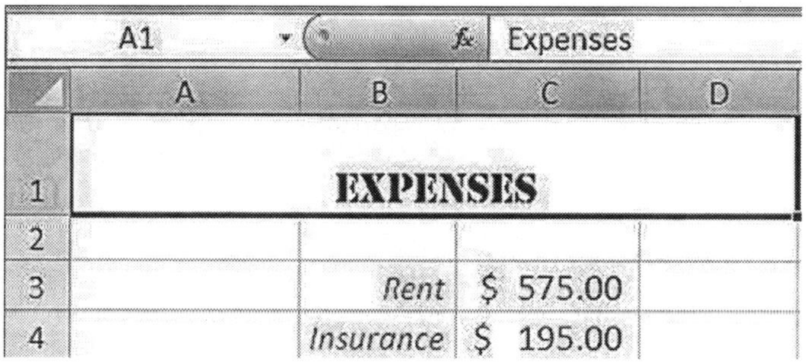

Figure 6-23

The last thing we want to look at in this lesson is the Wrap Text button.

Change the width of column A to 7.30

With the width severely reduced, the word Income will no longer fit inside the cell. In a few seconds you will get to see what Word Wrap can do.

Select cell A14 and then click on the Word Wrap button

The result is shown in Figure 6-24.

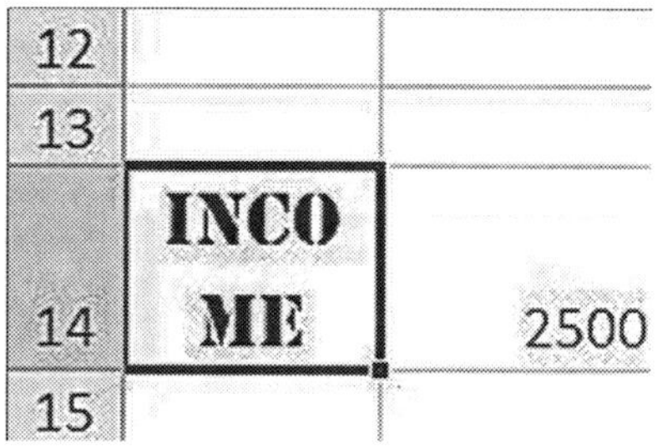

Figure 6-24

Granted this is not the way you would use the Word Wrap feature, but it does show what Excel can accomplish.

Click on cell D14 and then click the Word Wrap button

In the cell type Taco Bell Lunch date **and then click the Enter button** (see Figure 6-25)

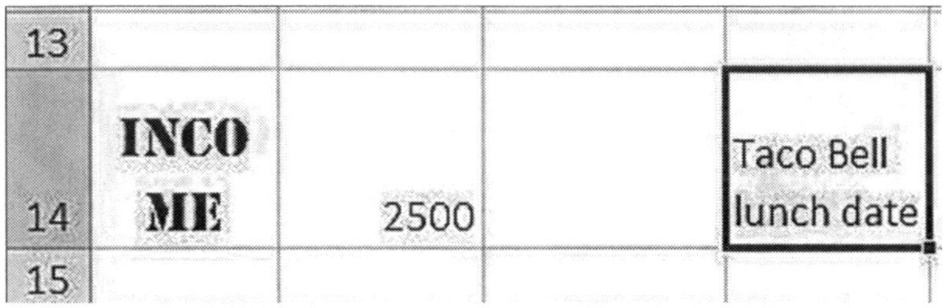

13				
14	**INCO ME**	2500		Taco Bell lunch date
15				

Figure 6-25

Word Wrap can be a great tool to use if you have text that will not fit inside one cell and there is no place for it to overflow into.

Note: Word Wrap will automatically adjust the height of the row to accommodate the text entered.

Lesson 6 – 6   Adding Borders

We have been looking at ways for you to spruce up your workbooks. Here is another way you can add that little something extra to your worksheets. You can add a border to a cell or even to a group of cells.

If necessary open the Income & Expenses workbook

Now would be a good time for us to total the monthly expenses.

Click on cell A11 and type Total Expenses **and then click the Enter button**

First things first, this label is in the same column as the other main spreadsheet labels. We should make this text look like the other labels.

Change the font and the size of the text in cell A11 to match the other labels

If you have forgotten what the font and size is for the other cells, simply click on the cell and check the Home tab of the Ribbon to see what it is. Now you will notice that the text is too large for the cell.

Click on cell A11 and then click on the Word Wrap command

Excel will put the text on two lines and show all of the text inside the cell.

Now we can add a formula to get the total expenses.

Click on cell D11 and then click on the AutoSum button

In the cell the following formula will appear: =Sum() with the insertion point flashing inside the parentheses. Excel doesn't know what numbers you want to add together, so we have to finish the formula.

Click on cell C3 and then using the keyboard type the colon sign (:) and then click on cell C10

Your formula should look like the one in Figure 6-26.

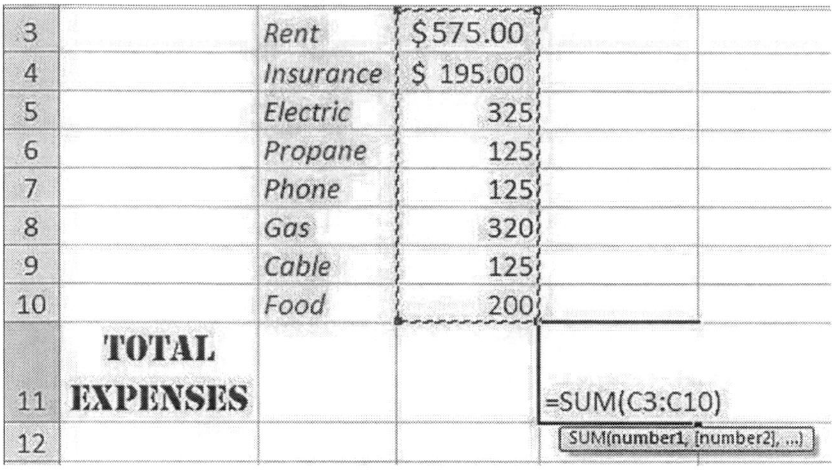

3
Rent
$575.00
4
Insurance
$ 195.00
5
Electric
325
6
Propane
125
7
Phone
125
8
Gas
320
9
Cable
125
10
Food
200
11
TOTAL EXPENSES
=SUM(C3:C10)
SUM(number1, [number2], ...)
12

Figure 6-26

Click the Enter button

Now that we have a total for the monthly expenses, we want this cell to stand out from the others. You should also note that the total is a currency because some of the entries were in currency.

With cell D11 still selected click on the down arrow of the Border command in the Font Group of the Home Tab (see Figure 6-27)

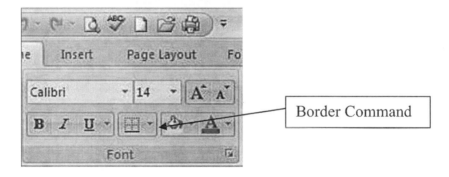

Insert
Page Layout
Calibri
14
Font
Border Command

Figure 6-27

When you click on the down arrow the following menu will pop onto the screen.

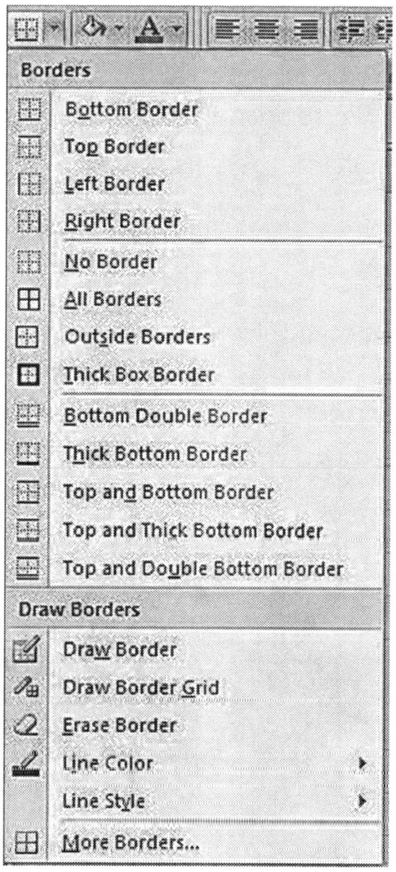

Figure 6-28

From this menu you can choose what type of border you want to add. You can add to the top, bottom, right and left sides, all sides, thick lines, double lines, almost any border that you want. You can even choose what color you want the border to be by choosing line color. You can also choose the style of the border such as dashes or dots.

Click on the All Borders choice and then click any other cell so you can see the border

Adding borders will help draw attention to important cells.

Save your changes

Lesson 6 – 7   Adding Colors and Patterns

Adding colors and patterns to your cells will also allow you to make a more attractive spreadsheet. Let's continue with our Income & Expenses workbook by adding a color and a pattern to some cells.

If necessary open the Income & Expenses workbook

Click on cell D11

We will add some background effects to this cell.

Click on the Dialog Box Launcher in the Font Group and then select the Fill Tab

The Format Cells Dialog Box is shown in Figure 6-29.

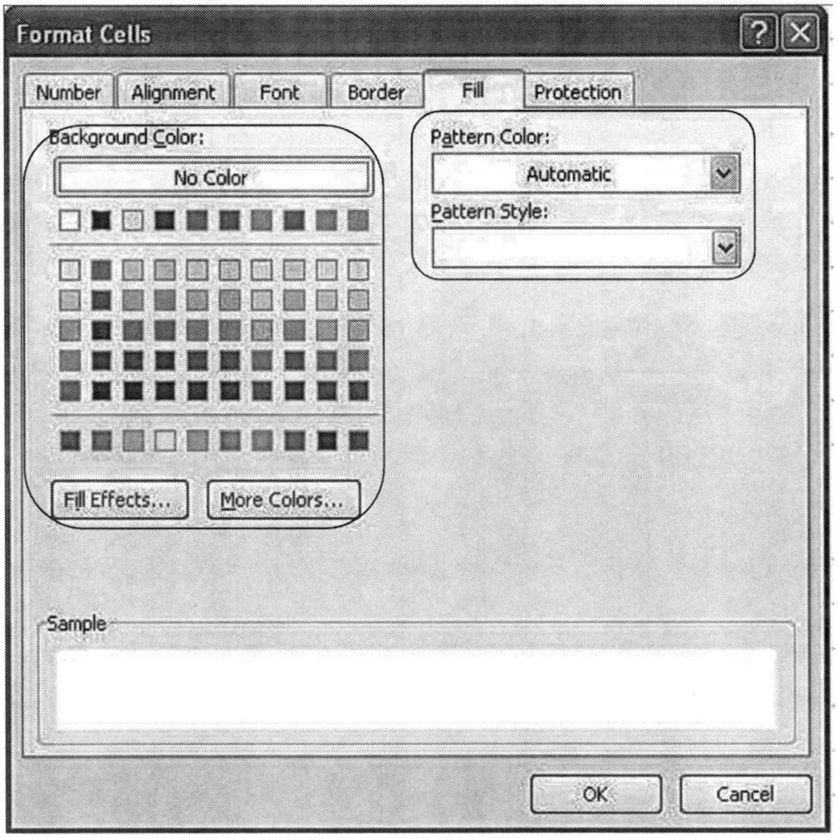

Figure 6-29

There are two sections to this dialog box. The first is the Background color located on the left side and the Fill Patterns on the right. The left side will determine the background color(s) and how it fills the cell. The right side will determine if there is to be a pattern displayed inside the cell. If you only want one basic color in the background, simply click on a color and then click the OK button. We want to do a little more than just that, so let's experiment.

Click on the Fill Effects button

Note: The "Fill Effects" and the "Background Color and Pattern" are mutually exclusive and cannot both be used in one cell.

The Fill Effects Dialog Box will come to the screen (see Figure 6-30).

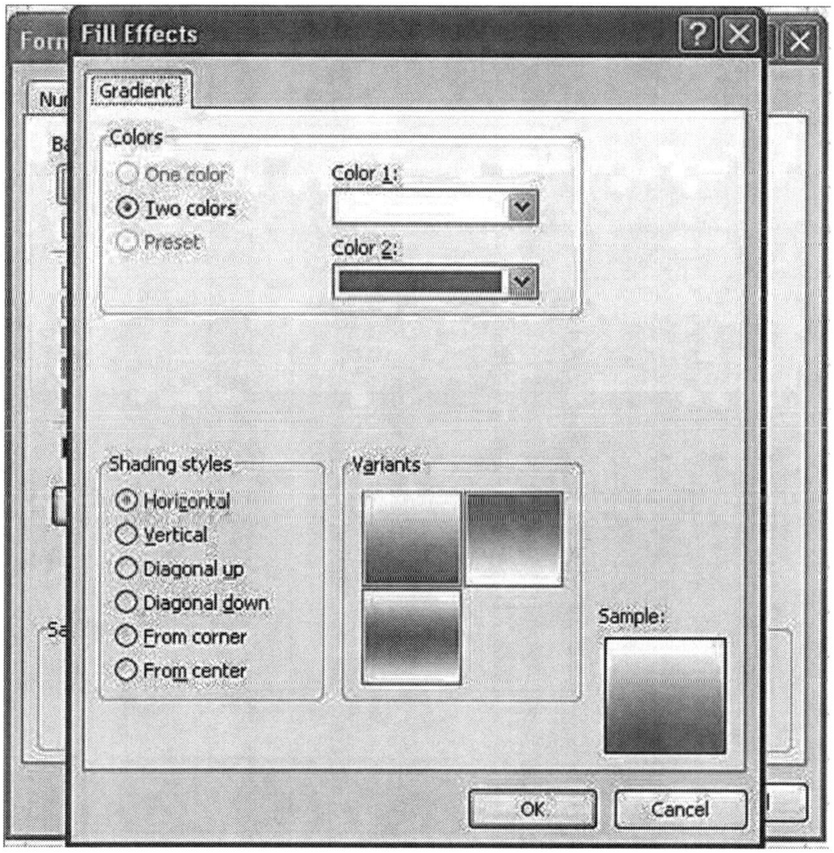

Figure 6-30

Click the various choices for both the Color 1 and the Color 2 and the Shading styles while watching the Sample area to see what the choices will look like

When you find a combination of colors and shading that you like, click the OK button

Just so you know, I am going to leave the default colors and the "From Center" shading.

Since the background color and pattern and the fill effects cannot both be used in one cell, let's select a different cell to try the pattern out.

Select cell A11 and launch the Font Dialog box again

Using the choice in the drop down lists make your dialog box look like the one in Figure 6-31

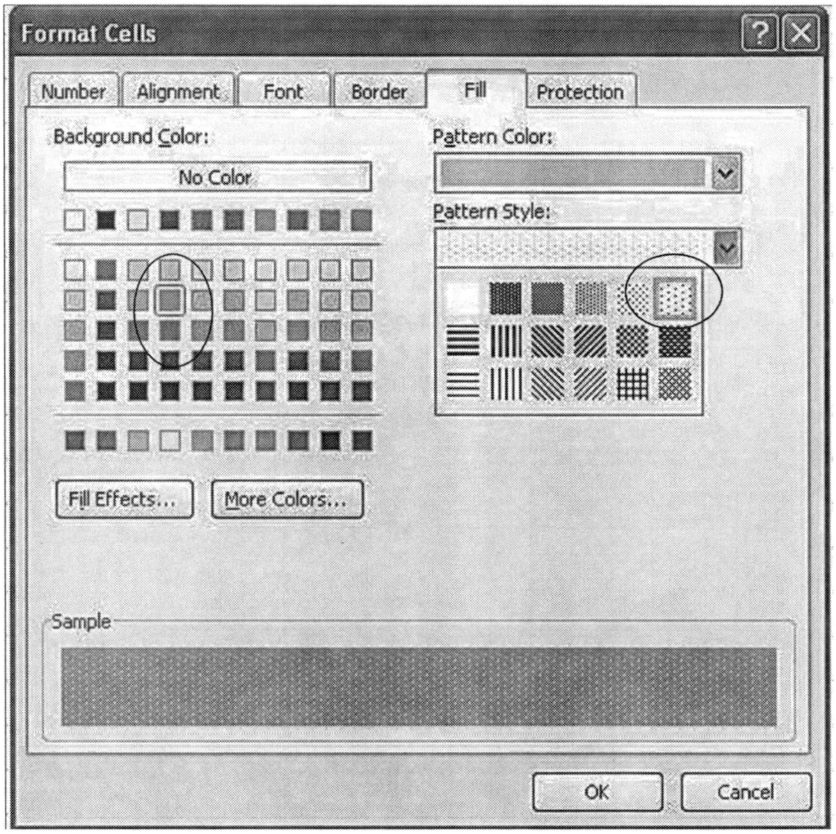

Figure 6-31

Click the OK button

What we did in this lesson can be applied to any cell or group of cells that we select. If you wanted, you could select the cells that show all of the expenses and the expense labels and change how these cells are filled. You are only going to be limited by your imagination.

Save your changes and close the workbook

Lesson 6 – 8   Using the Format Painter

We have covered everything that we are going to cover about how to format a cell. Now we are going to make the formatting a little easier. The Format painter will allow us to copy the format from one cell to another cell, with the click of a mouse button (well may be a couple of clicks).

The process for doing this is so simple that the entire lesson may not fill this page. The process involves three steps.

Click on a cell that is formatted the way you want it to be formatted.
Click on the Format Painter command button.
Click on the cell you want to copy the format to.

Click on cell D11

Click on the Format Painter button

Click on cell A14

All of the formatting that was in cell D11 is now copied to cell A14. The formatting from any cell can be copied to any other cell with only a few clicks of the mouse.

Chapter Six   Review

You can adjust the row height and the column width by clicking the mouse on one of the column or row boarders and dragging it to its new location.

Text can be formatted by changing the Font, the size, adding bold, italic, or underline, or by changing the color.

Values can be formatted by changing the style (such as currency or percent), adding commas and decimal points, or by changing the way it is displayed (including designing custom formats).

You can use Dialog boxes if a command is not visible on the Ribbon or if you need to make exact adjustments.

The text and values inside a cell can be aligned on either side of the cell or in the center of the cell. They can also be aligned with the top, middle, or bottom of the cell.

Word wrap will allow text that has too many words to fit on one line to be displayed on two or more lines inside a cell.

Two or more cells can be merged together and the text centered inside the one single cell. This is good for things like headings.

Borders can be added to any or all sides of a cell or group of cells to draw attention to them.

You can add colors and patterns to a cell or group of cells.

The Format Painter will allow you to quickly copy the formatting from one cell to another cell.

Chapter Six   Quiz

1) To adjust the width of a cell you need to double-click on the cell and fill in the desired width in the dialog box. **True or False**

2) To change the font type, you can type the name of the desired font in the Font Name box or choose the font from the drop down list. **True or False**

3) If a cell, or group of cells, need to be in new location, you must delete the entered information and retype it in the new location. **True or False**

4) Text color can be changed, but number colors must remain black. **True or False**

5) If you had the number 125 in a cell and then clicked the percent (%) command the result would be:
   a. 125%
   b. 1.25%
   c. 12500%
   d. 12.5%

6) To make a custom format to change a number from 123456 to 123-456, You would use which of the following:
   a. ###,###
   b. ***-***
   c. $3-$3
   d. ###-###

7) There is a command in the Alignment Group to allow the text to be printed at an angle. **True or False**

8) Word Wrap will automatically adjust the width of the cell to accommodate text that will not fit inside the cell. **True or False**

9) Borders can be put on all sides of a cell, except the top. **True or False**

10) Background colors and background patterns are mutually exclusive. **True or False**

Chapter Seven   Using Styles

If you want to have all of your worksheets look the same, you will love the Styles group of the Home tab. Appling styles to cells allows you to always have the same format every time you want to use it. Excel has a number of preset formats for you to choose from. You can also create your own style if you do not like any of the pre-made styles.

Lesson 7 – 1 Creating a New Style

If necessary open the Income & Expense workbook

In this lesson we will apply a style to some of the cells, we will also create a new style that we can use in the workbook.

Click on cell C5

We will use this cell to explore the pre-made styles.

Click the More button in the Styles Group of the Home Tab (see Figure 7-1)

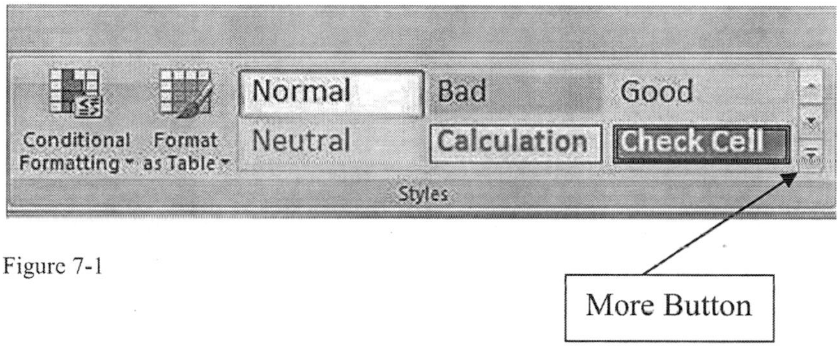

Figure 7-1

More Button

A new and scary dialog box will come onto the screen. We don't want anyone to have heart failure, so let's see if we can analyze it. This dialog box is shown in Figure 7-2.

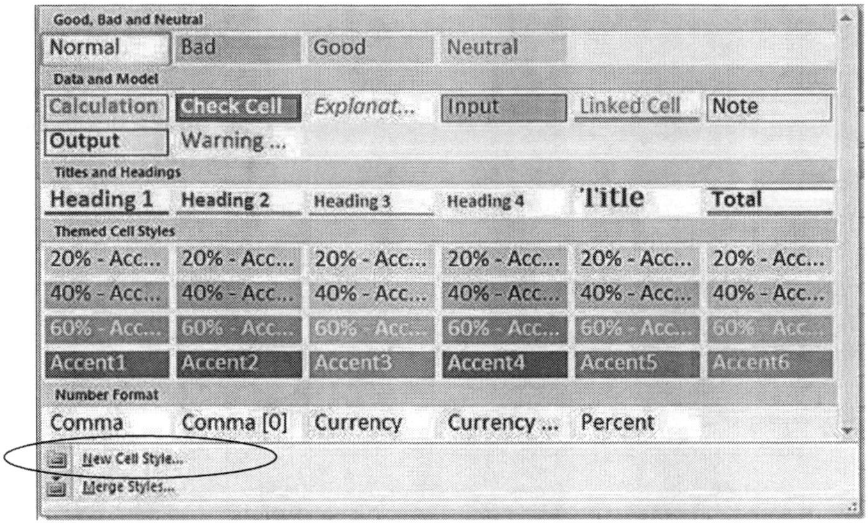

Figure 7-2

As you can see, there are many pre-made styles for you to choose from. Let's look at each one to see if there are any that you like.

Move your mouse over each style and watch cell C5 to see what the cell will look like with each style

To apply the style to the cell you just need to click on the style.

When you are finished, click on New Cell Style

Surprise, another dialog box jumps onto the screen. This dialog box is shown in Figure 7-3. We will be able to set the formatting that we want and give the style a unique name for future use.

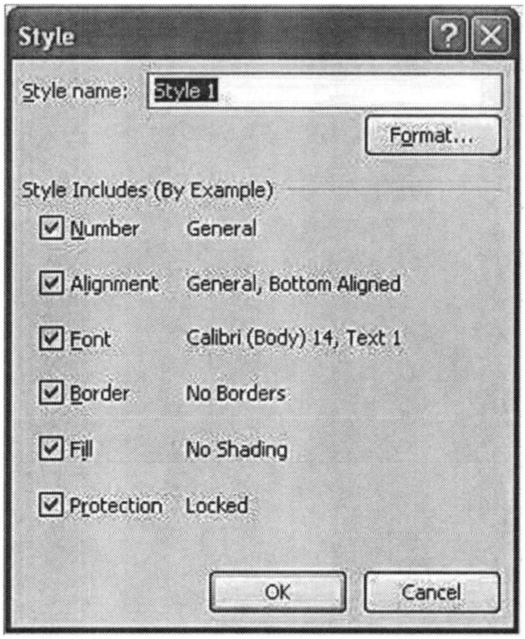

Figure 7-3

The first thing we need to do is to give the style a name.

In the Style Name textbox type the name Our Currency

After we give the style a name, we need to set the formatting for the style.

Click the Format button

Hey you have seen this before. It is the Format Cells Dialog Box. We can go to any tab we want and make any changes we want and then save it.

See if you can make these changes:

Change the number type to Currency
Change the Font to Arial
Change the background to light Gray

When you are finished click the OK button

The look of the cell should change to match the changes that you made.

Click OK on the Style dialog box

The next thing you need to notice is that the Styles Group has a new style in it. This style is called Our Currency. You can see this in Figure 7-4.

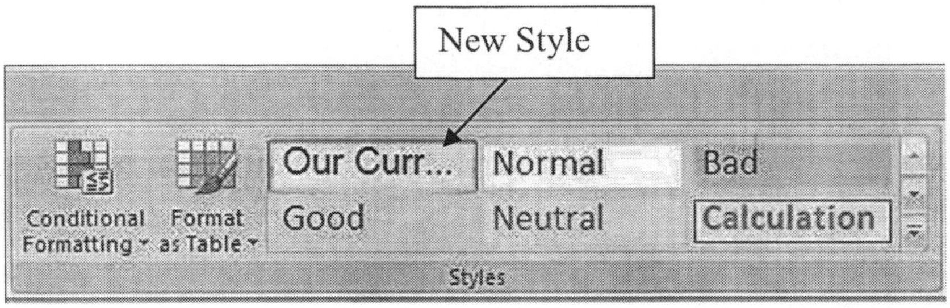

Figure 7-4

We will see how to apply this in lesson 7-3.

Lesson 7 – 2   Modifying a Style

There may come a time when you decide that you like a style, but you wish something was a little different. Excel gave us the ability to modify an existing style to meet our needs. In this lesson we will modify the style we just created in the last lesson.

Right Click on the Our Currency Style and select Modify from the drop down list

Wow, that was a surprise! This is the same dialog box we used to create the style. If you want, you can give this modified style a new name and keep the original as it is. You can also make the change and save it under the original name. If you do this the original style will be gone and it will be replaced by the modified style. Make sure this is what you want to do before you save it.

After you decide on the name, you can click the Format button and make any changes that you want. When you are finished, click the OK button to get back to the Style dialog box where you can click OK to save the changes.

Lesson 7 – 3   Applying a Style

This lesson has, for the most part, already been covered by lesson 7-1. However, I want to apply a style to some of the other cells in the Income & Expense workbook.

If necessary open the Income & Expenses workbook

In lesson 7-1 we created a new style called Our Currency. In this lesson we want to apply this style to the remaining cells under Expenses.

Click on cell C3

As soon as you click on cell C3, you will notice that Our Currency is no longer visible. It did not go away; it is just that the style for that cell is under the Currency style which is located at the bottom of the style list. Our Currency style is located at the top of the style list.

Click the Up arrow on the right side of the Styles Group box (see Figure 7-5)

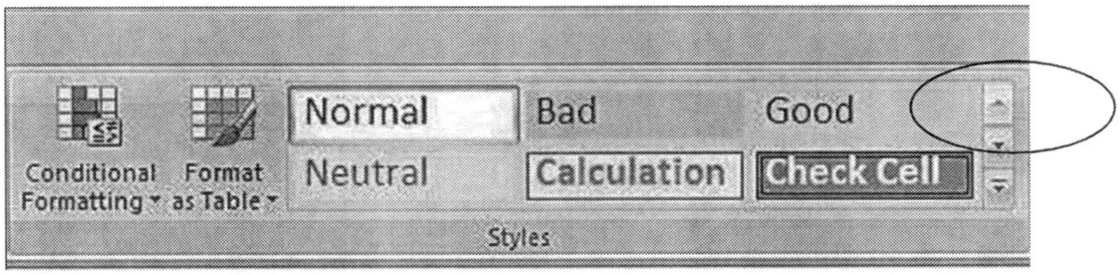

Figure 7-5

Continue to hold the mouse button down until the Our Currency Style is visible

Once the correct style is visible, we can apply it to a cell.

Click on the Our Currency style

The cell contents will immediately change to match the style we clicked on.

Repeat this process for cell C4

This style can be applied to more than one cell at a time.

Select cells C6 through C10

Click the Our Currency style button

The cells will immediately reflect the changes. Right now we have saved a style that is available for this workbook. If we want this style to be available for all new workbooks that we create, we must do a little more work.

Listed below are the steps to add this style to all new workbooks.

Make sure the workbook with the desired style is open
Open a new blank workbook by clicking on the Microsoft Office Button, the click New, then double-click on Blank Workbook
In the styles group click on the more button on the right side of the group
Click Merge Styles at the bottom of the dialog box
In the Merge style from box, click the workbook that contains the style(s) you want to copy, then click OK
Click the Microsoft Office Button, and then click Save As
In the File Name box type Book
In the Save As Type box click Excel template (you will probably have to use the down scroll arrow to find it)
In the box where you specify where you want to save the file, locate and select the XLSTART folder.
In Vista this is typically located in:
 C:\Users\user name\AppData\Local\Microsoft\Excel\XLSTART folder.
In XP the XLFOLDER is typically located in:
C:\Document and Settings\user name\Application Data\Microsoft\Excel\XLSTART
Click Save

Following these steps will allow the new style to be available for all new workbooks.

Chapter Seven   Review

Styles will allow you to have your worksheets look the same.

To create a new style, click the mouse on the "More Styles" button in the Styles Group. Then click the New Style command that is toward the bottom and on the left side of the Dialog Box. Give the style a unique name and then make the formatting changes needed.

A style can be modified by right-clicking on the style name that is displayed in the Styles Group and choosing "Modify" from the drop down list. You can also rename the modified style and keep the original style.

To apply a style, simply select the cell(s) you want the style applied to, and then click on the desired style.

Create a new style called My Style.

Format the style so the text is:

Times New Roman
Size 14
Font Color – Red
Background color – Light Blue

Type the following text in the cell: Good Style

Chapter Eight   Using Charts

Having a chart in your spreadsheet will give you one more item in your arsenal which allows you to have extraordinary spreadsheets. As you know a chart will allow you to display information in a graphical view. As you will see in this chapter, charts will have a tremendous impact on the way people view your spreadsheets.

Lesson 8 – 1   Creating Charts

Charts are very easy to create and only take a few seconds. Before you create a chart, you must have the data available. The easiest way to do this is to put the data into the spreadsheet. That is how we will start this lesson.

Open a new blank workbook

Enter the data as shown in Figure 8-1

	A	B	C	D	E
1	Sales By Region	North East	MidWest	North West	Total
2	Qtr 1	211704	87784	123094	422582
3	Qtr 2	172987	122514	108732	404233
4	Qtr 3	191485	143596	154689	489770
5	Qtr 4	225567	157630	174567	557764
6					

Figure 8-1

We need to select the data that is going to be part of the chart.

Select cells A1 through E5

Now we need to insert a chart.

Find the Chart Group on the Insert Tab and click on column (see Figure 8-2)

Figure 8-2

When you click on the Columns choice, you will get a dialog box that allows you to choose what type of column chart to insert. This is shown in Figure 8-3

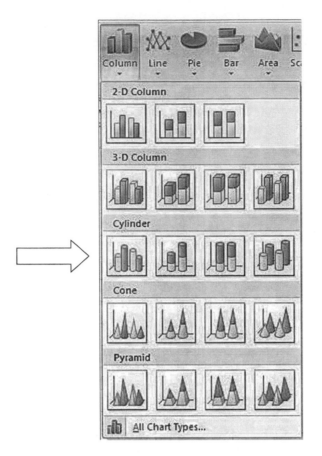

Figure 8-3

Click on the Cylinder solid color column chart (the first one under Cylinder).

As soon as you click on the type, the chart will jump onto the screen. As a matter of fact it will probably be in the center of the spreadsheet and that may not be where you want the chart positioned. Let's move the chart to a new location.

Before we move the chart, you need to understand that the chart has two sections. There is an inside section and an outside section. **Each section can be move independently.** The outside section will move the entire chart, while the inside section only moves the center part of the chart. The inner section may have to be moved to keep the chart legend from being placed too close to the columns. The chart also has sizing handles on the top, bottom, each side, and in each corner. These are represented by four small dotes placed together. Clicking on any of these will allow you to drag the size of the chart to make it smaller or larger.

228

Click the mouse between the numbers on the left and the outside edge of the chart and then drag the chart over to cell F1 (see Figure 8-4 and 8-5)

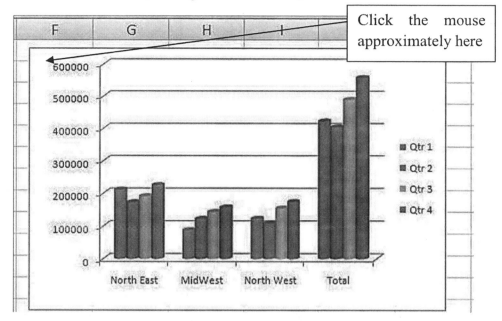

Figure 8-4

	A	B	C	D	E
1	Sales By Region	North East	MidWest	North West	Total
2	Qtr 1	211704	87784	123094	422582
3	Qtr 2	172987	122514	108732	404233
4	Qtr 3	191485	143596	154689	489770
5	Qtr 4	225567	157630	174567	557764
6					

Figure 8-5

When the chart reaches the desired location, release the left mouse button

As you can see the Region is shown on the bottom and each quarter is shown on the right, the height of the cylinder shows the amount of the sales.

The data is the same in both the spreadsheet layout and the chart layout. It is usually easier to grasp the concept if you see the data in the form of a graph.

You may not have noticed but there are three new tabs are on the Ribbon. These tabs deal with charts and are not normally visible. They become visible when they can be used, such as clicking on the chart. These tabs contain commands that are used when working with charts.

229

On the Design Tab and in the Data Group is a command that has Switch Row/Column on it. Clicking this button will switch the way the data is displayed. The Regions will move to the right side and the Quarters will move to the bottom.

With the chart still selected, click the Switch Row/Column button

The way the data is displayed changes as shown in Figure 8-6. You may have to adjust the size of the chart and move the center section to the left for yours to look like the one in the figure.

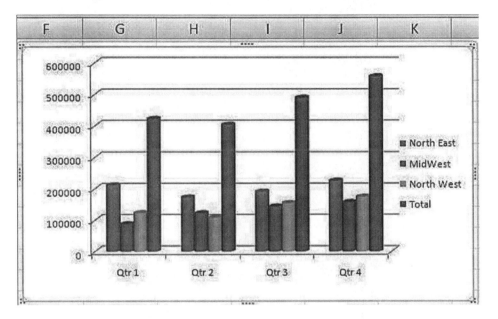

Figure 8-6

That is all there is to creating a chart. In the next lesson we will add to the chart.

Save your work as Sample Chart

Lesson 8 – 2   Adding Titles to the Chart

Having a chart in your spreadsheet is fantastic, but it might help if there were some titles to let the people know what the chart represents. In this lesson we will add titles to our chart.

If necessary open the Sample Chart workbook

Do you remember in the last lesson when I told you that you may have to resize the chart and also move the center section around to make room so things were not crowded together? Well now we get to try some of these things out.

Click on the center right sizing handle (see Figure 8-7) and resize the chart to make it about an inch wider

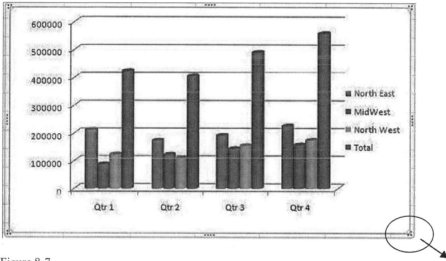

Figure 8-7

Now that the chart is wider, let's add some titles to help the viewer process this information.

Click on the Layout Tab of the Ribbon

Note: If the Layout tab is not visible, click on the chart and it will appear. This is one of the tabs that are visible only when it is able to be used. We will be looking at the Labels Group, in particular the Titles and Legends buttons. Figure 8-8 shows the Labels Group.

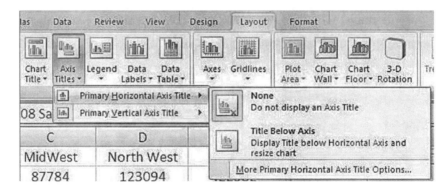

Figure 8-8

The first title that we will add is the Chart Title. This does just what the name implies; it provides a title for the chart.

Click on Chart Title and then select Above Chart

A textbox will appear centered and by the top of the chart. The default text is "Chart Title" and will need to be changed to something more appropriate.

Using the keyboard type 2008 Sales **and then press Enter**

The title for the chart will change to reflect your typing. Now that we have a title, we need to let the people know what the numbers on the left represent.

Click on Axis Titles and then Primary Horizontal Title and then Title Below Axis (see Figure 8-9)

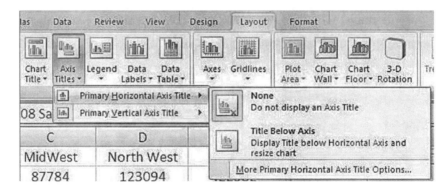

Figure 8-9

This will allow us to put a title under the chart telling the viewer what this part of the chart represents.

Using the keyboard type Sales by Quarter **and then press Enter**

The focus is already set so that when you start typing you will automatically be typing the title of the horizontal axis.

Repeat this process for the Vertical axis title and choose Vertical Title

Type US Dollars **for the title and then press Enter**

Your chart should look like the one in Figure 8-10.

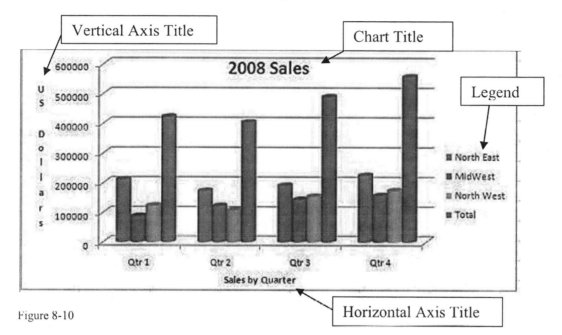

Figure 8-10

The titles can be changed at any time by clicking on them and retyping the name. For example, if you decide to click the switch row/column button you might want to change the horizontal title to Sales by Region.

If you decide that you don't want the legend, it can be removed or moved to another location. Clicking on the Legend button will allow you to remove it or move it.

Save your changes and close the workbook

Lesson 8 – 3   Using an External Data Source

This lesson will deal with using an external data source with a chart. In this lesson we will be using data from an Access database. The first thing we have to do is get the database into our spreadsheet and then we can display it in a chart. The database is included on the website for you to download or you may have copied it to your Excel 2007 folder.

Create a new blank workbook

Click on cell A1

Click the From Access command which is on the Data tab and in the Get External Data Group (see Figure 8-11)

Figure 8-11

Excel does not know where the data source is on your computer, so you have to tell it. As you have probably guessed, a dialog box will come onto the screen so you can tell Excel where the database file is stored. The Select Data Source Dialog Box is shown in Figure 8-12.

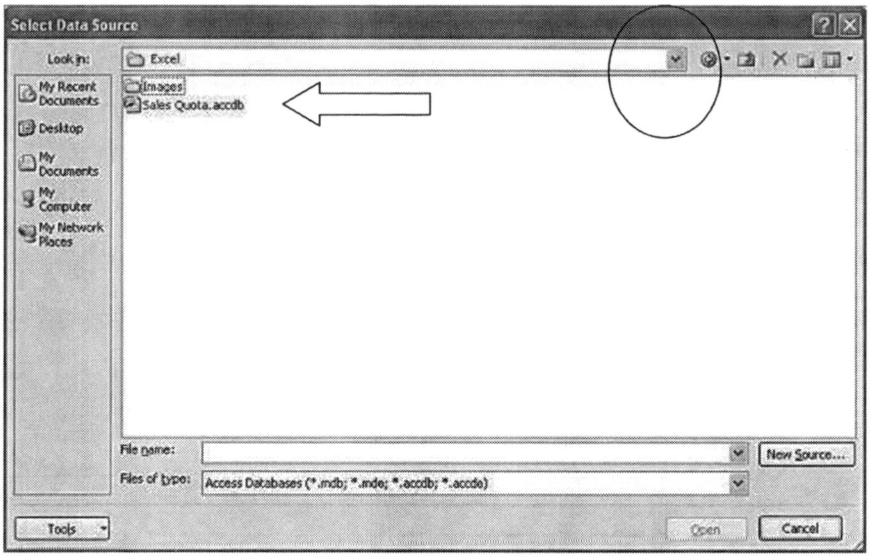

Figure 8-12

Using the drop down arrow, navigate to and select the Sales Quota database then Open

This will be found with the files that you downloaded or it may be in the Excel 2007 folder if you copied it to this folder.

Once you open this file there is another choice to make. The choice is where you want the file placed in the spreadsheet. The Import Data dialog box will allow you to make this choice (see Figure 8-13).

Figure 8-13

235

We want to view the data in a table and we want it to be in the existing worksheet. The main choice is which cell we want to start with. The default cell A1 is already in the dialog box. Notice that this is an absolute reference which is indicated by the dollar signs. This is where we want the data to be inserted, so all we have to do is click OK.

Click the OK button

As soon as you click the OK button, the data will jump onto the screen as shown in Figure 8-14.

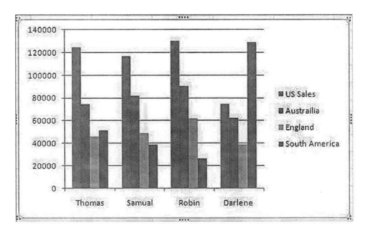

Figure 8-14

Now that the information is in the spreadsheet, we can insert a chart to display it.

Using the method we used in the last lesson, insert a chart to display this data (do not include the ID in the chart) and use the Clustered Column type of chart

Your chart should look like the one in Figure 8-15

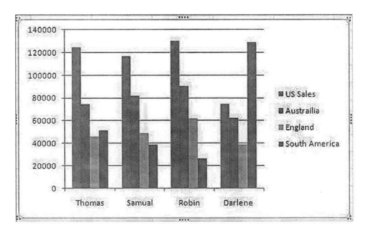

Figure 8-15

If you have trouble, go back and review the previous lesson.

The chart will show each salesperson's sales for each country. Just for fun, let's compare each salesperson's totals compared to the other salespersons.

Click the Switch Row/Column button to see how the salespersons compare to each other

As you can see, Darlene had fewer sales in every marketplace except South America, which she clearly dominated. This is much easier to see with this view than the previous view.

Now we have a potential problem. What happens if, on your way to the boardroom where you have to present your data to the sales manager, someone goes into the database and changes the data to reflect the new data that has just become available?

Note: You cannot make a change to the database while it is open in the spreadsheet. You will need to close it in the spreadsheet before you can make any changes.

In the Connections Group of the Data Tab, there is a button that has "Refresh All" on it. Clicking this button will update the contents of the spreadsheet to match the database. The change will immediately be reflected in the chart.

Save the workbook as Sales Quota and close it

Lesson 8 – 4 Using Pie Charts

The charts we have been using so far work just great for comparing the data that we have been comparing. Some data, however, is better seen using a different kind of chart. I am referring to a pie chart. This type of chart is great for comparing trends and relationships.

Create a new blank workbook

Fill in the data until it looks like Figure 8-16

A8

	A	B	C
1	Media	New Customer Analysis	
2	Newspaper Add	32	
3	Phone Book	8	
4	Web Site	26	
5	Friend	19	
6	Signage	8	
7	Other	7	
8			

Figure 8-16

This data represents a survey from new customers defining how they heard of our business. We want to display this information in a chart that will represent 100 % of our new customers.

Select the cell range from 2A to B7

Insert a 3-D pie chart (see Figure 8-17)

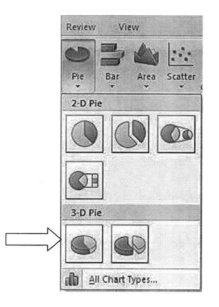

Figure 8-17

This chart will show what percentage of the whole each part represents and will look like Figure 8-18.

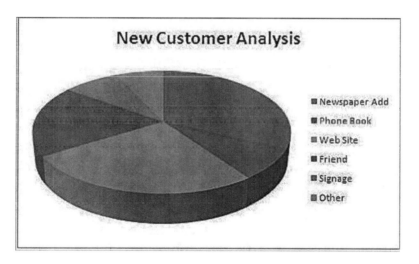

Figure 8-18

This is not all you can do with a pie chart. If you click on the pie chart and move the mouse over each section of the chart you will see the value of each section (see Figure 8-19).

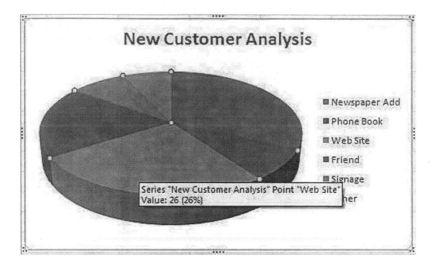

Figure 8-19

This is not all that you can do either. You can pull the pie apart and show each section separated from the others. If you click inside the pie chart once, you can see dots at each joint as shown in Figure 8-19. If you click a second time on one section the dots for that section will remain and the others will go away. If there are no dots showing on the pie chart and you click one any piece and drag it away from the others, the entire pie will separate. If the dots for only one section are showing and you click and drag the section away, the rest of the pie will remain intact. This is shown in Figures 8-20A and B.

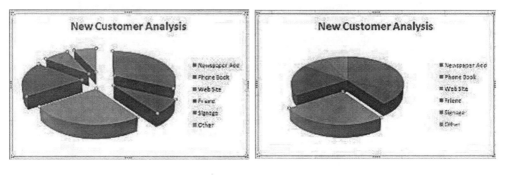

Figure 8-20A   Figure 8-20B

Try separating the pie chart

Note: You can always use the Undo to put the pie chart back together.

Save your workbook using the name Survey

Lesson 8 – 5   Styles and Effects

When you make a chart it looks pretty good like it is, but you can dress it up a little and give it a little pizzazz. You do this with the fill effects.

Open the Survey workbook if necessary

Click on the chart to select it

To get to the different effects, we need to use the Format tab of the Ribbon. In particular we need the Shape Styles Group. This group is shown in Figure 8-21.

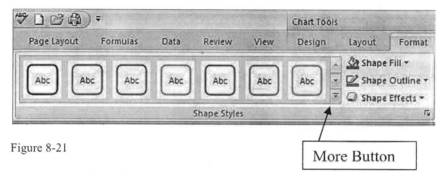

Figure 8-21

More Button

Move your mouse over each of the shape styles

You will be able to see what the chart will look like with the chosen style as you pass over it. You do not have to select a style and apply it before you can see how it looks on your chart. As you can see, these only change the outside border color. If you find one you like you can click on it to select it. If you do not see one you like, you can click on the More button to see other styles.

Click the More Button

A new menu of styles will drop down full of styles for you to consider. This is shown in Figure 8-22.

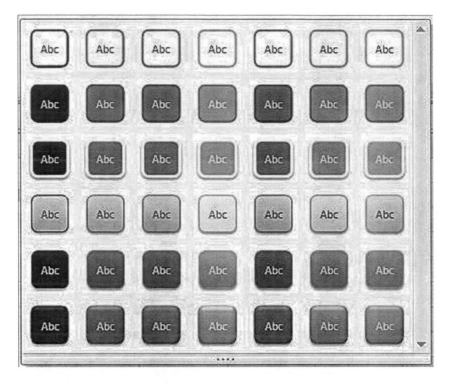

Figure 8-22

If you choose one of these styles it will change more than just the border around the chart. These styles will also change the background color and the text color. Again if you move the mouse over these choices the change will occur when the mouse moves over the style so you can see what the change will look like. If you see the style you want, simply click on it to make the change.

Move your mouse over each choice and view each of then

When you are finished, click anywhere outside of the choices to remove this box from the screen

There are three other choices you can consider in this group. The shape fill command will allow you to change the background color of the chart. The shape outline command lets you change the border color. The shape effects will allow you to change a few other things. You need to try this to understand it. I can't explain it well enough.

With the chart selected, click on the shape effects and move the mouse over all of the choices to see the effect they have on the chart

Save and close the workbook

242

Chapter Eight   Review

Charts will let you display information in a graphical view.

To create a chart you must first have the data in the worksheet and selected. You can then click on the chart type in the Chart Group on the Insert Tab.

Charts can be resized and moved to show the data better.

Titles in the charts can make it easier to understand the data. You can have titles for the entire chart, the horizontal axis, and the vertical axis.

An external data source can be imported to the worksheet and then used in a chart. The external data can also be refreshed if the data changes.

Pie charts show 100% of the data in the form of a pie. All of the pieces of the pie can be separated or a single piece can be pulled away from the rest of the pie.

You can use styles and effects to dress up your chart.

1) Create a pie chart showing the following data:

	A	B
1	Region	Visitors
2	Northeast	129000
3	Southeast	216000
4	North Midwest	473000
5	South Midwest	460000
6	Desert States	80000
7	Northwest	68000
8	Southwest	72000

2) Create a column chart showing the following data:

	A	B	C	D	E	F
1	ID	Sales Person	US Sales	Austrailia	England	South America
2	1	James	147000	84000	51000	52000
3	2	William	115000	78000	55000	38000
4	3	Robin	129000	91000	68000	25000
5	4	Diane	82000	63000	46000	130000

Do not include the Employee ID in the chart.

Chapter Nine   Using Tables

As you already know, a worksheet, or spreadsheet, consists of cells that are organized into columns and rows. In this worksheet you can create a table to manage and analyze data completely independent of any data that is outside of the table. (In previous versions of Excel a table was called a list)

Lesson 9 – 1   Creating a Table

Open a new blank workbook

Insert the following data into the worksheet:

	A	B	C	D	E
1	Product	Qtr 1	Qtr 2	Qtr 3	Qtr 4
2	Peanuts	2307	2416	2300	2509
3	Mint Patty	1408	1299	1504	1616
4	Almond Joy	1499	1518	1567	1700
5	Snickers	1600	1705	1800	1654
6	Hershey Bar	1708	1815	1987	1899
7	Crackers	1415	1345	1488	1504
8	M & M	1463	1609	1548	1623
9					

Figure 9-1

The first thing we have to do is select the cells we want in our table. These cells can be empty cells or they can already have data in them.

Select cells A1 through E8

After we select the cells we need to format them as a table.

On the Home Tab and in the Styles Group click on the Format as Table command (See Figure 9-2)

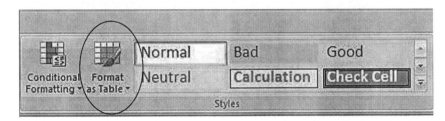

Figure 9-2

Clicking on this button will bring a chart full of styles to the screen as shown in Figure 9-3.

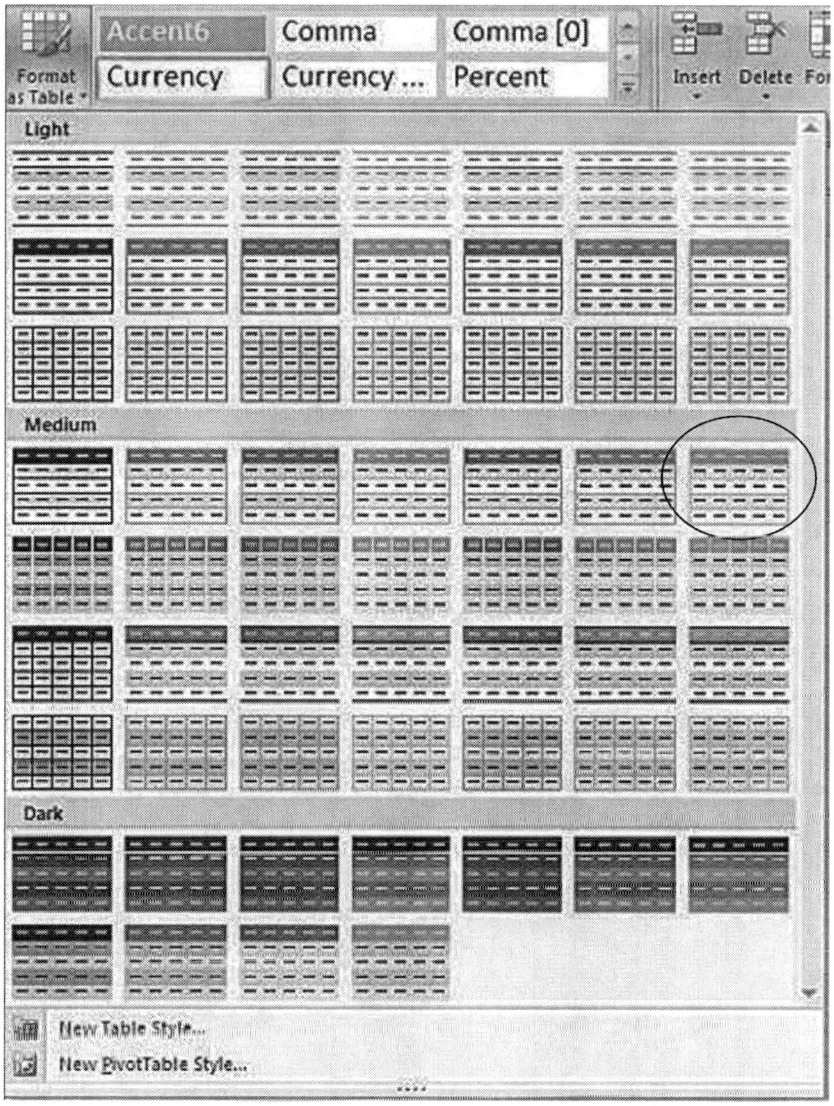

Figure 9-3

This will allow us to pick a style that will give our table a professional look and feel. For our choice let's chose the top right choice under medium.

Click on the top medium orange choice

We now have one thing left and that is to confirm that we want to format these cells as a table. Our last choice is shown in Figure 9-4.

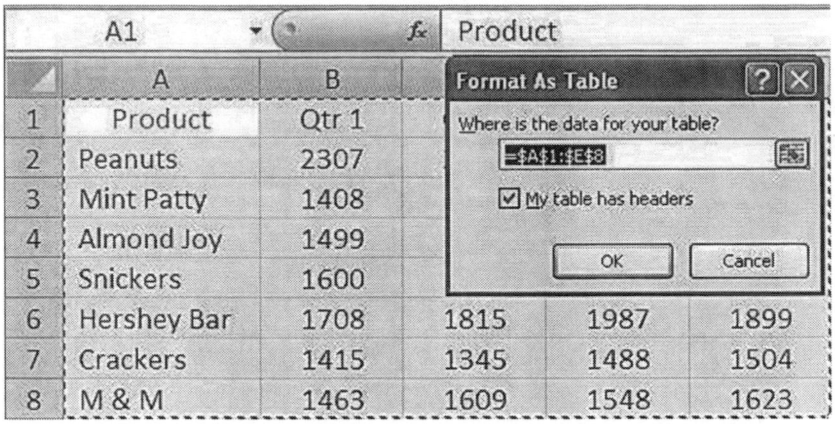

Figure 9-4

The cells that we have selected are shown in the "Where is the data for you table" section. There is one other part you will want to notice. There is a check box to determine if there are headers for your table. Since there are headers across the top, we want this box checked.

Click the OK button

The finished product is shown in Figure 9-5.

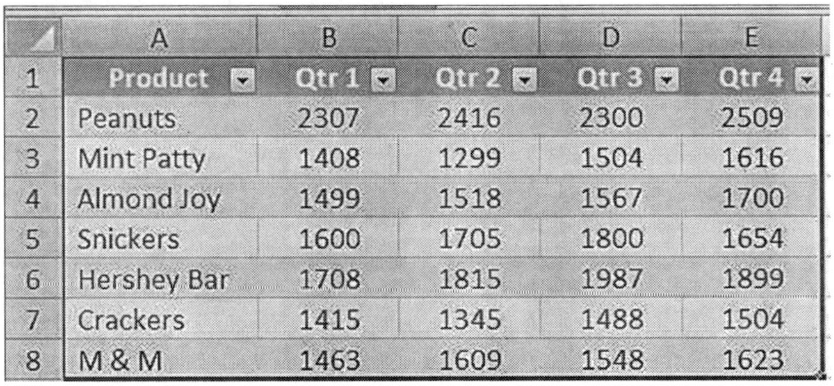

	A	B	C	D	E
1	Product	Qtr 1	Qtr 2	Qtr 3	Qtr 4
2	Peanuts	2307	2416	2300	2509
3	Mint Patty	1408	1299	1504	1616
4	Almond Joy	1499	1518	1567	1700
5	Snickers	1600	1705	1800	1654
6	Hershey Bar	1708	1815	1987	1899
7	Crackers	1415	1345	1488	1504
8	M & M	1463	1609	1548	1623

Figure 9-5

Every other line is highlighted to allow the data to stand out and be more easily identified. You will also notice that there is drop down arrow by each header label. These will be discussed more in the next lesson.

Note: You can also click on the Table command on the Insert Tab to insert a table.

The next lesson will show us some of the things we can do with a table.

Save your work as Table_Sample

Lesson 9 – 2   Working With Data in a Table

Now that we have our table created and data in it, we need to know how to work with the data. In this lesson we will see how to add to the table, do calculations using the table, and what happens when we change the data in the table.

Open Table_Sample **if necessary**

The first thing we will want to do with our table is add a Totals column.

Click on cell F1 and type Grand Total **and then press Enter**

You will see that the column is included in the table and the words Grand Total are considered the header for the column. Now we will need something in this column.

With cell F2 highlighted, click the AutoSum button

The results are shown in Figure 9-6.

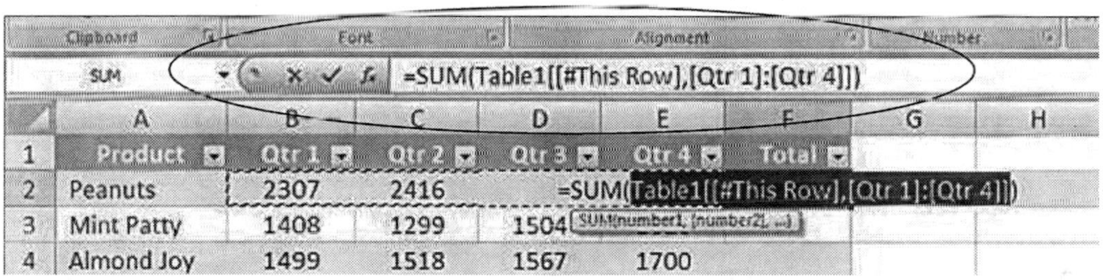

Figure 9-6

There are a few things you need to be made aware of in this illustration. As with all formulas, the equals sign is at the beginning. The next is that it is a sum. The next part identifies which table is being used in the formula. Since this is the first table in the worksheet it is Table 1. Next we are taking the numerical values in this row, the one the cell is in, and then it lists the starting and stopping cells listed by the header name not the cell name.

Click the Enter button on the Formula bar

When you click the Enter button something neat will happen. All of the cells below this will automatically filled in with the formula. You don't have to copy the formula or use the fill handle to have the formula added for each of the other cells. If you click on each cell in the Grand Total column you will notice that the formula does not change. This formula will work for any row because it does not reference individual cell names.

We have added a column that will give us totals for each row, now let's add a row that will give us totals for each column. We don't even have to select a row or a cell for Excel to know where to put the Totals row. It will automatically be put below the last row.

You might have noticed that there is a new Table Tool section to the Ribbon. On it is the Design Tab. This tab contains tools that you use to work on tables.

On the Design Tab click on the Total Row in the Table Style Options Group (See Figure 9-7)

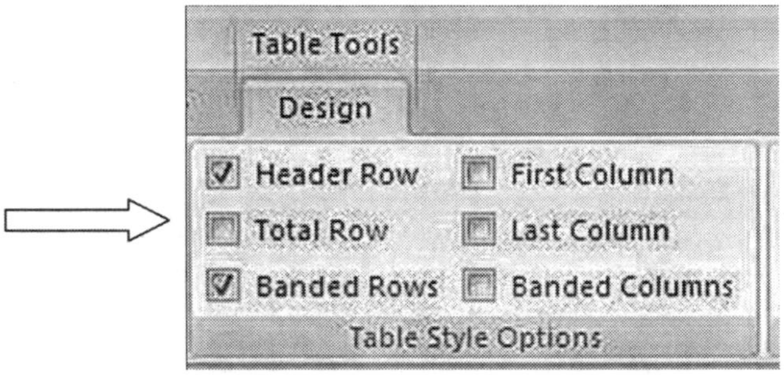

Figure 9-7

When you click on the Total Row command, the Total Row will be inserted automatically. Now we have several options on how and what we want in the Total Row.

Select cell B9 and then click the down arrow on the cell (see Figure 9-8)

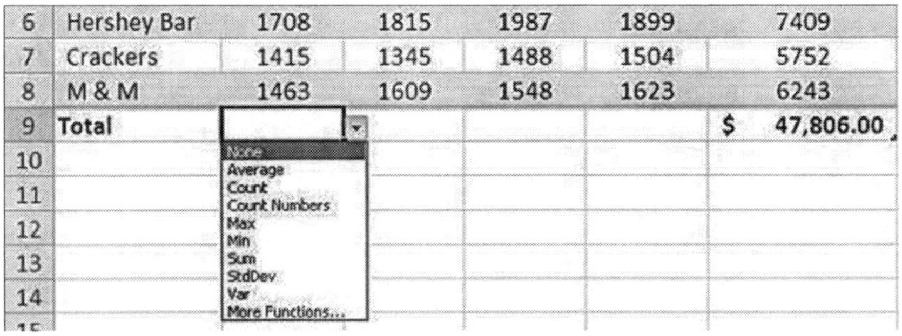

6	Hershey Bar	1708	1815	1987	1899	7409
7	Crackers	1415	1345	1488	1504	5752
8	M & M	1463	1609	1548	1623	6243
9	Total					$ 47,806.00
10						
11						
12						
13						
14						

None
Average
Count
Count Numbers
Max
Min
Sum
StdDev
Var
More Functions...

Figure 9-8

From here we can choose to have our total show several different things. For our choice we are going to use the sum.

Click on the Sum choice

Repeat this for the rest of the cells in the Totals Row (except the last one that is already filled in)

The filled in cells are shown in Figure 9-9.

	A	B	C	D	E	F
					=SUBTOTAL(109,[Qtr 4])	
1	Product	Qtr 1	Qtr 2	Qtr 3	Qtr 4	Grand Total
2	Peanuts	2307	2416	2300	2509	9532
3	Mint Patty	1408	1299	1504	1616	5827
4	Almond Joy	1499	1518	1567	1700	6284
5	Snickers	1600	1705	1800	1654	6759
6	Hershey Bar	1708	1815	1987	1899	7409
7	Crackers	1415	1345	1488	1504	5752
8	M & M	1463	1609	1548	1623	6243
9	Total	11400	11707	12194	12505	47,806.00

Figure 9-9

A new feature of Excel 2007 is called Structured Reference. This lets you reference a group of cells by referencing its header name. The next part you will not readily recognize is the 109 in the formula. This refers to the function number. This is not something you will know, but Excel knows what it means. If you want to know what the function numbers are, you can click the help button (the small blue question mark on the top right) and enter function number in the search box.

Note: If you want to add a row to the end of your table, you can click on the last cell in the last row and then press the Tab key. A new row will be inserted into the table and the totals will reflect the additional data that is added in the new row.

Note: you can also insert a row above an existing row by right-clicking on any cell of the row and then selecting insert and Table Rows above from the shortcut menu.

Now let's add another column to show the average number of items sold.

Click on cell G1 and type Average Items **and then press Enter**

You may have to adjust the column width to see all of the words in the header (you could also click the Word Wrap button).

251

With cell G2 selected click the down arrow on the AutoSum button

Select the Average Choice

When you click on Average, the default values will be in the formula. You will notice that this includes the Grand Total column. This was included because it is a number and Excel doesn't know that we don't want it in the average. Now we need to change what cells are included in the average.

Click on cell B2 and drag the mouse over to include cell E2

The formula will now look like Figure 9-10.

	A	B	C	D	E	F	G	H
	SUM				fx	=AVERAGE(Table1[[#This Row],[Qtr 1]:[Qtr 4]])		
1	Product	Qtr 1	Qtr 2	Qtr 3	Qtr 4	Grand Total	Average Items	
2	Peanuts	2307	2416	2300	2509	=AVERAGE(Table1[[#This Row],[Qtr 1]:[Qtr 4]])		
3	Mint Patty	1408	1299	1504	1616	AVERAGE(number1, [number2], ...)		
4	Almond Joy	1499	1518	1567	1700	6284		
5	Snickers	1600	1705	1800	1654	6759		

Figure 9-10

Click the Enter button

The remaining cells going down will be automatically filled in for you and each will show the average for quarters 1 through 4.

Right-click on cell G9 and select Average

You might think that this will follow the same pattern as the cells above it and give the average of the cells to the left. Since this is a Totals Row and gives the total for the column, this will also give the average of the column.

The finished product should look like Figure 9-11.

	A	B	C	D	E	F	G
1	Product	Qtr 1	Qtr 2	Qtr 3	Qtr 4	Grand Total	Average Items
2	Peanuts	2307	2416	2300	2509	9532	2383
3	Mint Patty	1408	1299	1504	1616	5827	1456.75
4	Almond Joy	1499	1518	1567	1700	6284	1571
5	Snickers	1600	1705	1800	1654	6759	1689.75
6	Hershey Bar	1708	1815	1987	1899	7409	1852.25
7	Crackers	1415	1345	1488	1504	5752	1438
8	M & M	1463	1609	1548	1623	6243	1560.75
9	Total	11400	11707	12194	12505	$ 47,806.00	1707.357143

Figure 9-11

You will probably like the layout of a table, the way it looks with its banded rows. You may like it so much that you may want to use it even if it is not a table. You can convert a table back to a range of cells by clicking on the Convert to Range command in the Tools group of the Design tab. The formatting will stay the same but the data will no longer be inside a table.

Save your work

Lesson 9 – 3   Creating a PivotTable Report

If you are looking for a fast and powerful way to analyze data, then the PivotTable Report is the tool you want to use. You can use a PivotTable report to analyze and summarize your data. With PivotTable reports, you can look at the same information in different ways with just a few mouse clicks.

Let's start with some of the general rules you will need to understand.

The worksheet you are using must be well prepared.
Each column must contain similar data types. That means that you should have text in one column, numbers in another column, and dates in another column. Do not mix data types in the column. If the column contains text do not put anything but text in the column.
There can be no empty columns.
There can be no blank rows, such as a blank line to separate sections of data.
Each column must have a title or label at the top at the top of the column. Each column will become a field you can use in the report. The name for each field will come from the title at the top of the column.

Open the workbook titled Sales Data

There is enough data in this spreadsheet to allow us to see how a PivotTable Report can help us analyze data.

You can use all of the data that is available or you can select only the data you want to use. For our report we are going to use all of the data.

Click the PivotTable command that is in the Tables group of the Insert tab

The Create PivotTable Dialog box will open. The dialog box is shown in Figure 9-12.

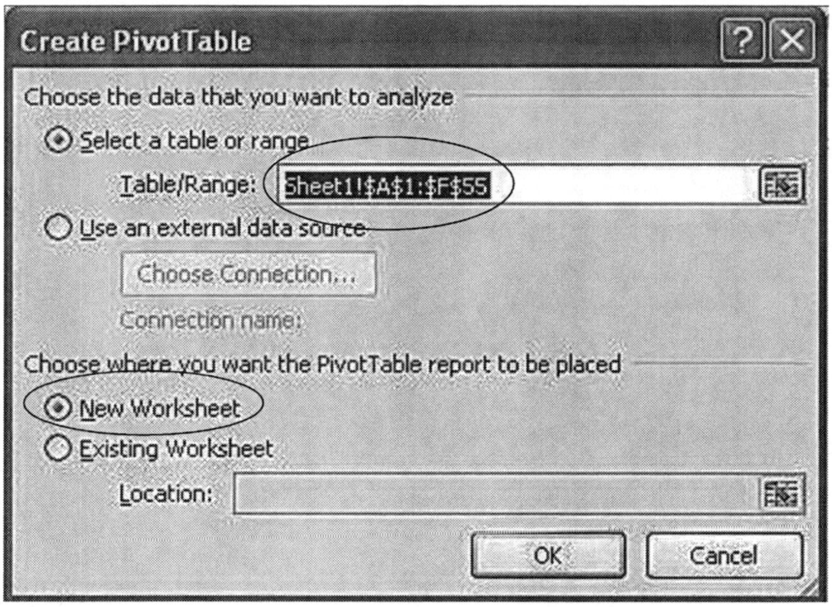

Figure 9-12

The select table or range is already filled in for you and includes all of the data in the worksheet. You will notice that you will be using Sheet 1. There is also an absolute reference starting with cell A1 and ending with cell F55. You will also note that the PivotTable Report will be placed in a new worksheet. You could place it on the same worksheet as the data, but I prefer putting it on a new sheet.

Click the OK button

As soon as you click the OK button a new worksheet will be on the screen, and is shown in Figure 9-13. The center part of the worksheet has been deleted from the images so you can see both sides.

The layout area for the PivotTable report   The PivotTable Field List

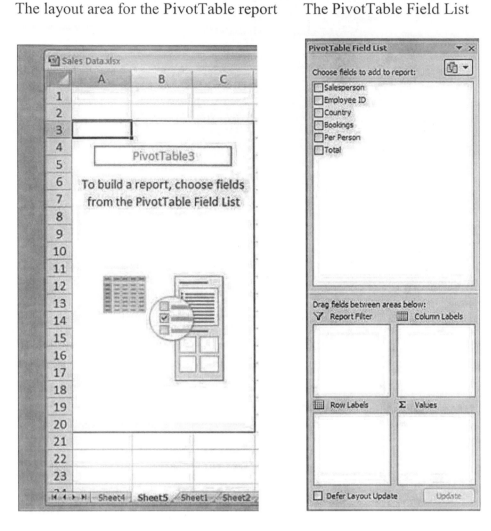

Figure 9-13

On the left side is the layout area and on the right side are the fields. We now must select the fields we want in our report. Not every field needs to be in a report. Some reports may require the Employee ID to be included, others may not.

In our report, we want to see the number of trips that were booked by each employee and the revenue that was brought in by each employee. It should be obvious that we will need the Salespersons in our report, so we will need to click the checkbox next to Salespersons.

Click the checkbox next to Salespersons

256

The Salespersons column has text in it so it is automatically put on the left side of the report and in the first column. Even if you did not check this box first, Excel would have put it in the first column. Text always goes in the first column.

All of the names of the Salespersons appear on the left side of the report layout area.

Next we want to list the trips by country.

Click the checkbox next to Country

Each country, since it also contains text and not numbers, will be nested under each salesperson.

Next we want to know how many bookings each salesperson sold.

Click the checkbox next to Bookings

You will notice that there is a total across from each name and each country has its own total.

The last thing we need in our report is how much money each salesperson brought into the company.

Click the checkbox next to Total

The grand total of the sales are listed for each salesperson and it is broken down further by each country.

Now we can look at our report and get the answers to our questions about the bookings and income produced by our salespersons.

If you want to change the order of the data in the report, such as see the total revenue generated before the number of bookings, the PivotTable report will allow us to make changes. At the bottom right, in the field's list side, are four rectangles that we can work with. These are shown in Figure 9-14.

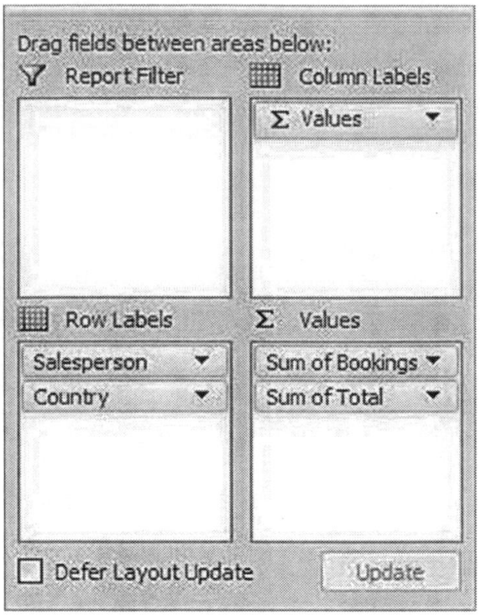

Figure 9-14

If you want to switch the Sum of Bookings with the Sum of Total, you could click on one of them and drag it either up or down to the desired location. You could also right-click on one of them and select to move it up or down. This is shown in Figure 9-15.

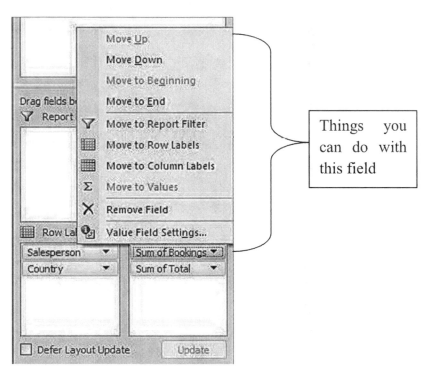

Things you can do with this field

Figure 9-15

You will notice that there is a minus sign next to each salesperson's name. If you click this, the detail will disappear and you will only have the totals. The minus sign will be replaced by a plus sign, which, when clicked, will bring the detail back to the screen. This is shown in Figure 9-16.

	Row Labels	Sum of Bookings	Sum of Total
3		**Values**	
4	**Row Labels**	**Sum of Bookings**	**Sum of Total**
5	⊟ **Diane**	185	124385
6	Canada	37	18463
7	France	21	14679
8	Italy	51	38199
9	Spain	16	10384
10	Sweeden	30	26970
11	UK	30	15690
12	⊞ **Mary**	295	191727
13	⊞ **Melissa**	284	162386
14	⊞ **Patty**	189	133661
15	⊞ **Robert**	268	177914
16	⊞ **Thomas**	145	92205
17	⊞ **William**	253	154175
18	**Grand Total**	1619	1036453

Figure 9-16

Using the PivotTable Report can easily and quickly give you answers, and help you analyze your data. This is definitely something you should consider using.

Save your changes

Lesson 9 – 4   Specifying the PivotTable Data Source

The PivotTable Report is great, but what if the data you need is not in the workbook? Microsoft thought of this also.

Open a new blank workbook

Click PivotTable on the Insert tab

The Create PivotTable Dialog box will come to the screen. This time there is nothing in the Select Table or Range text box. This makes sense since there is nothing in the worksheet for it to assume you are going to use. This time we need to choose the data source.

Click the radio button next to choose an external data source and then click choose connection

As usual a dialog box will pop onto the screen. This one is the Existing Connections Dialog box and shows existing network connections that you can choose from. If there is an existing connection and it is the one you use to get tō a database or spreadsheet, you can click on it. There are no connections on my computer, so I am going to have to browse for my data source.

Click on the Browse for More button

The Select Data Source Dialog box will jump onto the screen. The data source we want is not in the My Data Sources folder, so we will have to navigate to the Excel 2007 folder (or the folder with the downloaded files) and then choose the Sales Quote.accdb file.

Navigate to the Sales Quota.accdb file and then click the Open button

Click OK on the Create PivotTable Dialog box

Now you are back at the PivotTable layout and Field List and ready to add the fields to the layout area.

That is all there is to using an external data source.

Chapter Nine   Review

A table will allow you to analyze and manage data independently from any data that is outside the table.

Select the cells that you want in the table and then format them as a table.

You can total rows and columns that are located inside the table.

If necessary, you can add rows and columns to a table.

A PivotTable Report will allow you to analyze and summarize data. Remember there are certain rules about not mixing data types that must be followed. Not all fields must be used in the PivotTable Report, you can choose to include only the ones you want in the report.

Chapter Nine   Quiz

1) Data outside of a table will not affect the data inside a table. **True or False**
2) Once you format cells as a table, they must remain a table. **True or False**
3) Table use relative references for the cells. **True or False**
4) The AutoSum command, when used in a table, uses a row number and a cell name in the formula. **True or False**
5) The "Total Row" command on the Design Tab, when checked, will cause a Total Row to be automatically inserted into the table below the last row in the table. **True or False**
6) If you want you can have the Total Row only show the largest number in the column. **True or False**
7) A PivotTable Report can distinguish between different data types, so it is okay to mix data types in a column. **True or False**
8) You can choose to put a PivotTable Report on a separate worksheet if you desire. **True or False**
9) It is mandatory that every filed available be included in the PivotTable Report. **True or False**
10) You can create a PivotTable Report even if the data used in it is not part of the workbook. **True or False**

Chapter Ten   Using Graphics

Inserting graphics and pictures into your worksheet will give your spreadsheets a little more pizzazz. That little extra will make your spreadsheets look like a real professional has created them. In this chapter you will learn how to insert pictures into your spreadsheet, resize the pictures, and move the pictures. You will also learn how to add a textbox to the spreadsheet.

Lesson 10 – 1 Inserting & Working with Pictures

Open a new blank workbook

Fill in the data until it looks like Figure 10-1

	A	B	C	D	E
1		Average Temperature & Percipitation			
2					
3		Qtr. 1 Avg.	Qtr. 2 Avg.	Qtr. 3 Avg.	Qtr. 4 Avg.
4	Avg. High	51.0	78.3	87.7	60.0
5	Avg. Low	23.7	49.7	59.7	33.3
6	Avg. Percip	2.7	4.2	3.6	3.5
7					

Figure 10-1

The spreadsheet shows the average high and low temperatures as well as the average precipitation per quarter. Let's add a picture to the spreadsheet and see the difference it makes.

First we will need to create a little space between the title and the actual data, so the picture will have a place to reside.

Select cells A3 through E7

To move the selected cells to a new location, you must move the mouse pointer to one of the borders where it will turn into a four-sided arrow and then click the left mouse button and drag the entire selection to the desired location, and then you need to release the mouse button.

Move the selected cells down five rows

This will leave a blank space for the picture to be placed.

Click on cell A2

This will give us a starting place for our picture. Now we need to select the picture we want to insert. The picture can be one we have made with a program such as paint, or it can be a picture we have downloaded from a camera or the web, or it can be from clip art. The picture we are going to use is one that was included with the download, and is called Clouds.jpg.

Click on the Picture command that is on the Insert tab

The Insert Picture Dialog box pops onto the screen as shown in Figure 10-2.

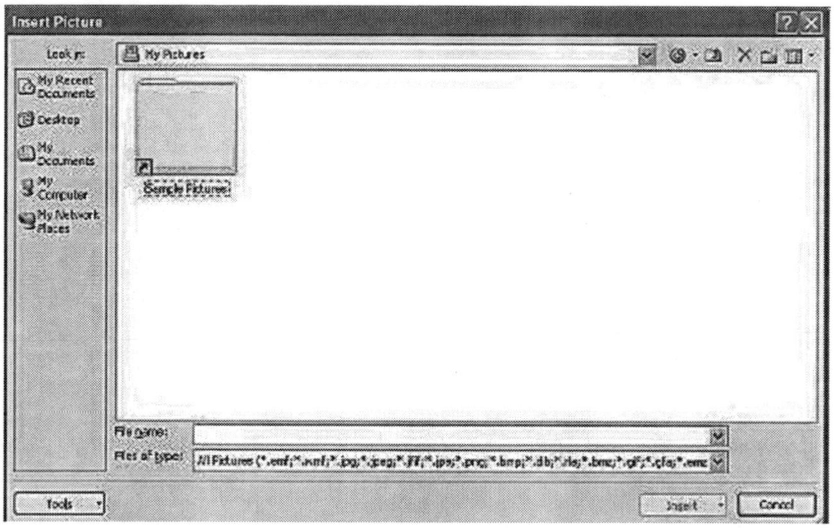

Figure 10-2

The picture that we want is not in the My Pictures folder, so we will have to navigate to it.

Navigate to and select the picture called Clouds.jpg

The picture can be found with the downloaded files or in the Excel 2007 folder, if you copied it to that folder.

Click on the Insert button

You will immediately notice that the picture is much too large to fit in the small space that we have provided for the picture.

Now we have to resize the picture to make it fit.

The picture (shown in Figure 10-3) has sizing handles on it. These are located along the sides and in the corners. The easiest way to resize a picture is to "grab" it by one of the sizing handles and drag it to the proper size. I suggest that you use the bottom right sizing handle this time. You will know when you are able to use the sizing handle because the mouse pointer will change to a double sided arrow. When it changes, click and hold the left mouse button and drag the mouse to the up and left at the same time. The picture will get smaller as you drag the mouse.

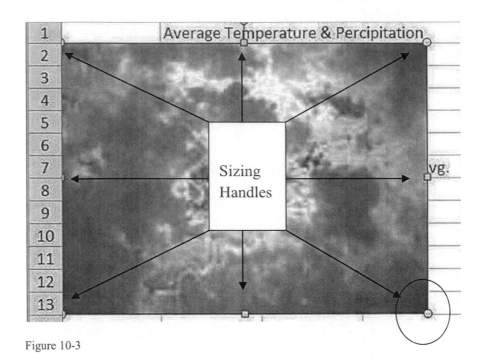

Figure 10-3

Resize the picture until it is the size of the one in Figure 10-4

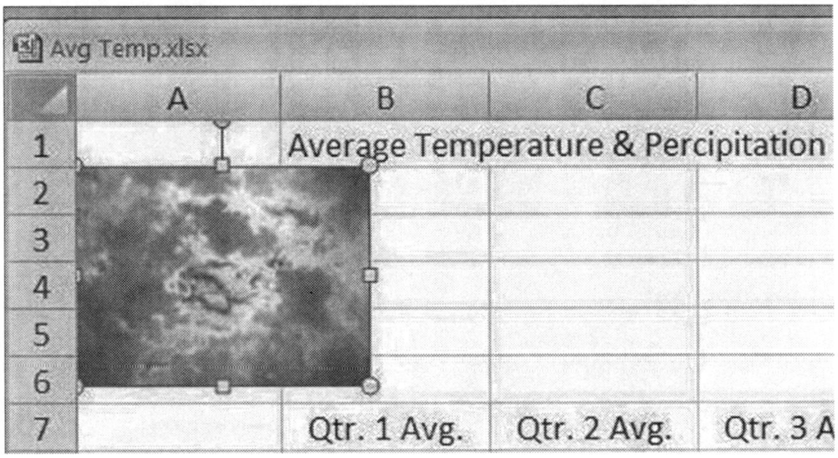

Figure 10-4

The picture is now small enough, but it doesn't look quite right. It would look better if we centered it under the title.

To center the picture we will have to move it from its current location.

Move the mouse pointer inside the picture

When it gets inside the picture it will change to a cross with arrows at each end. To move the picture all you have to do is click and hold the left mouse button down while you drag the picture to the desired location.

Move the picture until it is approximately where it is in Figure 10-5

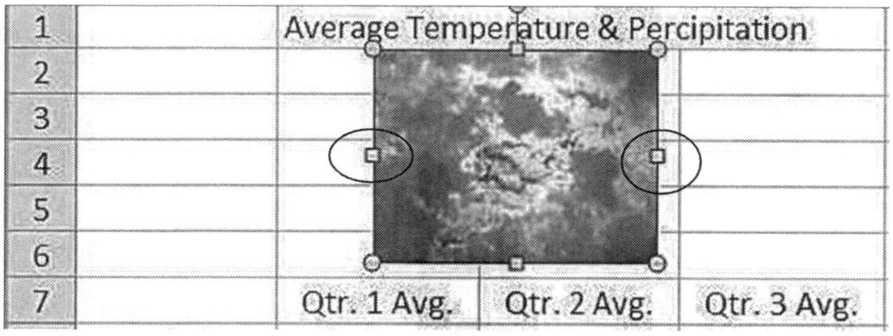

Figure 10-5

Now that we have the picture moved to a better location, we need to adjust the size one more time. This time we will change the width of the picture.

Using the left and right sizing handles change the width to match Figure 10-6

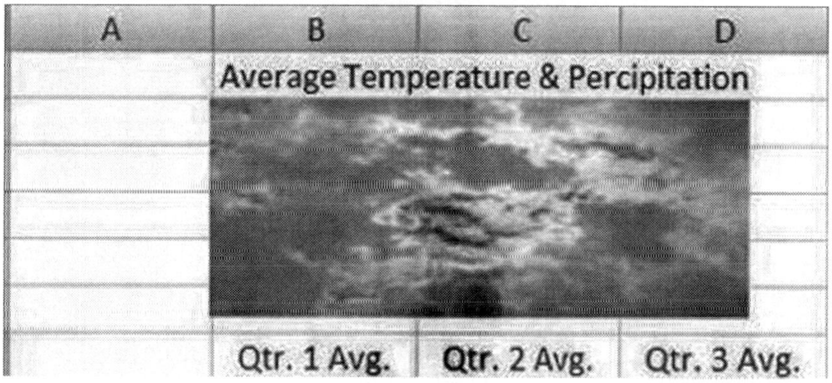

Figure 10-6

A picture can also have the height changed by using the resizing handles at the top and bottom. We won't need to change ours, but the height can be changed as easily as the width was changed.

Save the workbook as Avg Temp

Lesson 10 – 2   Using the Textbox

The Textbox can be a useful tool in your spreadsheet. Using the textbox you can add text to help the user understand what you are trying to convey to the viewer. You might want to think of the textbox in the same way you would a comment, only it is visible all the time. In this lesson we will add a textbox to the Avg Temp worksheet.

Open the Avg Temp workbook if necessary

Click the Textbox command on the Insert tab (See Figure 10-7)

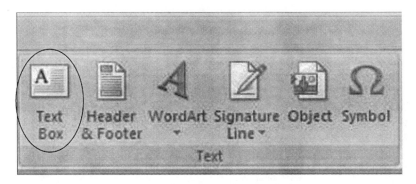

Figure 10-7

When the Textbox button is clicked, it will turn to amber and you will know that it is ready for you to draw your textbox, the mouse pointer will also turn into a thin upside down cross.

Move your mouse to cell F7 and click and hold the left mouse button, and then drag the mouse until it covers part of H10 or until it looks like Figure 10-8 and then release the left mouse button

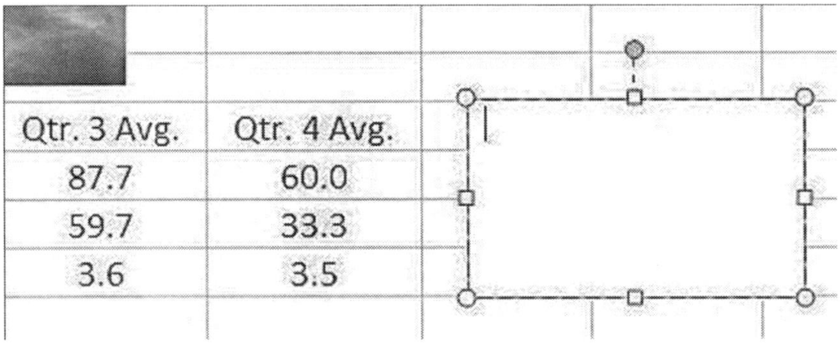

Qtr. 3 Avg.	Qtr. 4 Avg.
87.7	60.0
59.7	33.3
3.6	3.5

Figure 10-8

268

As you can see, the textbox will have the insertion point (the flashing vertical line) already in it ready for you to start typing.

Type the text in the textbox as shown in Figure 10-9

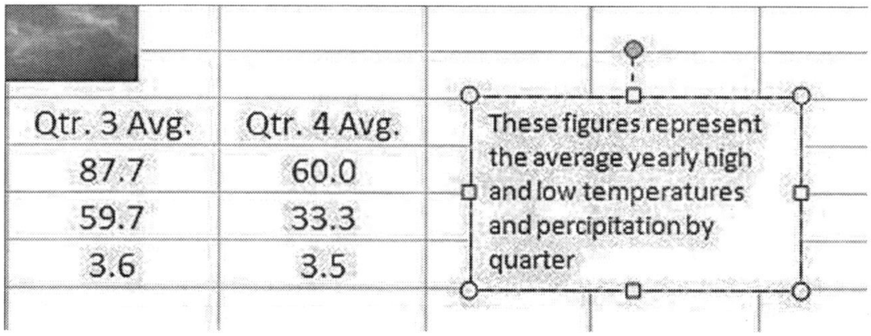

Qtr. 3 Avg.	Qtr. 4 Avg.	These figures represent
87.7	60.0	the average yearly high
59.7	33.3	and low temperatures and percipitation by
3.6	3.5	quarter

Figure 10-9

You do not have to worry about pressing the enter key at the end of every line. The textbox uses the word wrap feature, which means the text will automatically move to the next line as it comes to the end of one line.

To remove the dotted border around the textbox, click anywhere outside of the textbox. If you have trouble getting the text to fit correctly inside the textbox you can use the sizing handles to change the size. These work the same as the ones for the pictures in the last lesson.

Textboxes are easy to put into your worksheets, they arc a little fun, and they can help your viewer to make sense out of your spreadsheet.

Save your work

269

Chapter Ten   Review

Using pictures in your spreadsheets will add to your arsenal of tools for making your spreadsheets more appealing.

To add a picture click on the cell where you want the picture inserted and then click on the picture choice that is on the Insert Tab. Next locate the desired picture and then click the Insert button of the Dialog Box.

You can change the size of the picture or move it to a new location to better fit your needs.

Adding a Textbox will let you provide information to the person viewing the worksheet.

Chapter Ten   Quiz

1) To move a picture to a new location in the worksheet you use the "Move" command on the Ribbon. **True or False**

2) Using the _____ you can add text to help the user understand what they are seeing.

3) The sizing handle that is located in the middle of the right side of the picture will allow you to change the height and the width of the picture at one time. **True or False**

4) When the Textbox button on the Ribbon is clicked, the mouse pointer will change to an amber color. **True or False**

5) When entering text in a textbox, you must press the enter key at the end of every line. **True or False**

Chapter Eleven   Using Hyperlinks

A hyperlink is what it sounds like. It is a link to allow you to move to another place in a document or even another document. You have probably used these many times, even if you didn't know what they were called. Almost every website has a place that you can click on to move to another page or to another section in the same page.

Lesson 11 – 1 Creating a Hyperlink to a Specific Place in the Workbook

Open the Hyperlink Example workbook

This workbook will be with the downloaded files or in the Excel 2007 folder if you copied it there. Across the top of the worksheet is a row that has Move To on it. Each month's average is also listed. We are going to use these cells to move to the row with the monthly average for that month.

If you are going to use a hyperlink to move to a specific place in the workbook, the destination must have a specific name so the hyperlink can find it. Let's start by giving the end point a name.

Click on cell A9

After we select the cell we should define a name for the cell. We discussed how to name cells in Lesson 4 – 1.

(On the Formula Tab and in the Defined Names group is the Define Name command).

Click on the Define Name command

The New Name Dialog box will jump onto the screen (See Figure 11-1). The default name is the text that is in the cell. You can leave this or type in a new name. For this lesson let's change the name a little.

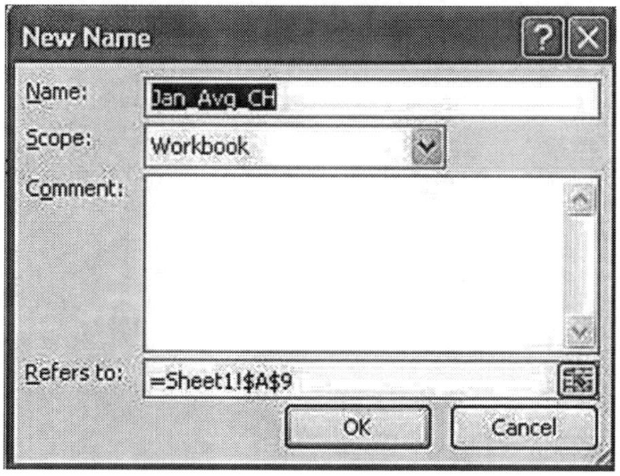

Figure 11-1

Type Jan_Avg **and then click the OK button**

If you just casually look at the worksheet, you may not notice any difference. Actually the cell name has changed as you can see in Figure 11-2.

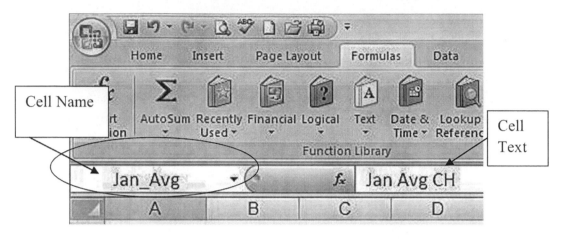

Figure 11-2

Now that the cell has a name, Excel can find it.

Note: The cell had a name before you changed it to Jan_Avg it was A9 and you could have used that instead of giving it a new name, but that wouldn't have been near as much fun.

Now we need to set up the hyperlink between cell B1 and cell A9.

Click on Cell B1

On the Insert tab and in the Links group, click on Hyperlink

Clicking the Hyperlink button will bring the Insert Hyperlink Dialog box to the screen (shown in Figure 11-3).

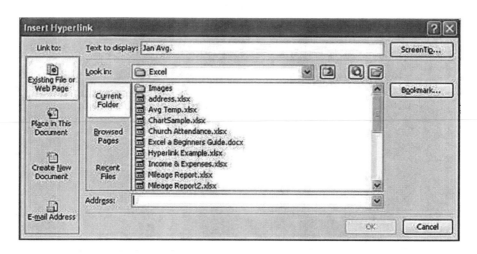

Figure 11-3

The dialog box is divided into three sections. On the left we tell Excel where the other end of the link is found. On the top we tell Excel what text to display in the cell that has the link. On the bottom we tell Excel exactly where the destination end of the link is located.

Top:

> The way our worksheet is laid out, default text will work fine for the hyperlink. We want to move to the January Average so the default text will convey that to the user.

Left Side:

> The default choice for where the general destination is located is another file or web page. This won't work for this hyperlink. The next choice is to link to a place in this document and that is what we want. The last two choices will allow us to link to a new document or have the link go to an e-mail address.

Right Side:

> The right side will allow us to browse to a new location if the link is not in this workbook. If the link is inside this workbook we need to use the Bookmark button to browse to the named cell.

Note: The Bookmark button and the button to link to a place in this document both have the same results.

Click on the Place in This Document button on the left side

275

The right side of the dialog box will change to let you enter the destination cell reference or use the defined name that you created. See Figure 11-4.

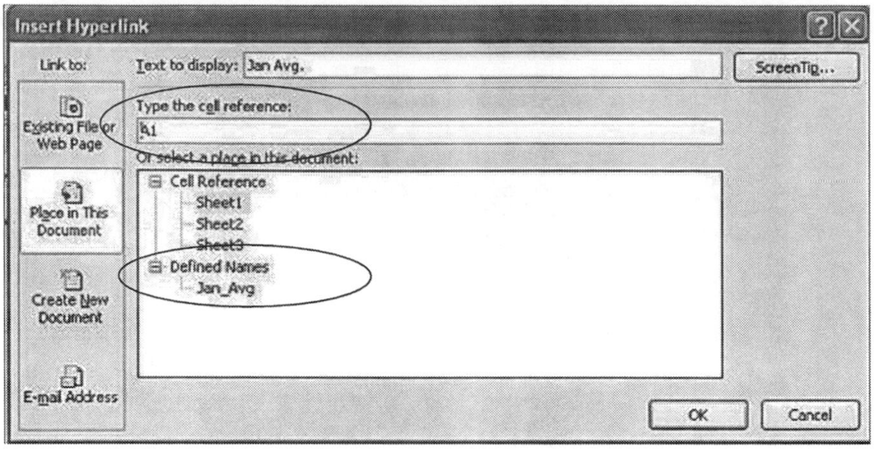

Figure 11-4

Click on the defined name Jan_Avg **then click OK**

You may not notice a lot of change, but the text in cell B1 has changed to blue and has a line under it (See Figure 11-5). This identifies it as a hyperlink.

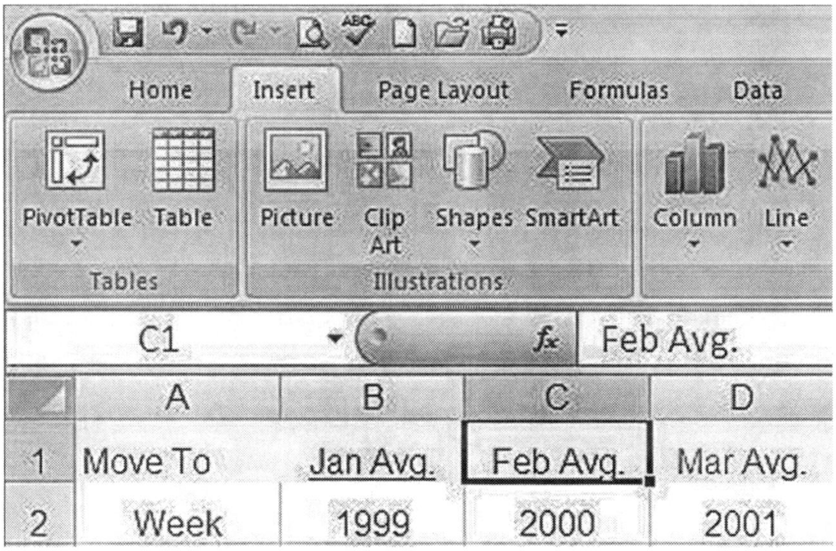

Figure 11-5

The hyperlink has been created, now all we have to do is see if it works.

Move the mouse to cell B1 and then click the mouse

When the mouse gets over the hyperlink the mouse pointer will change into a hand. That is when you need to click the mouse. The focus will change to cell A9 and it will now have a border around it.

Make hyperlinks for the remaining cells on the first row

Save your work

> **Note:** A hyperlink can be entered manually by using the formula =HYPERLINK("path to the destination file", cell name to go to). Example =HYPERLINK("C:\sales\Sales Data.xls'. A1)

Lesson 11 – 2 Creating a Hyperlink to another Workbook

Creating a hyperlink to another workbook is just as easy as creating one to a place in the existing workbook. This lesson will show you the differences between the two.

Open the Trips workbook

We will create a hyperlink from the workbook to the PivotTable report in the Sales Data workbook.

Click on cell H1 and then click on the Hyperlink command in the Links group of the Insert tab

In the Text to Display type View PivotTable Report

Click Existing File or Web Page if it is not selected

Navigate to and select Sales Data.xlsx that is either with the downloaded files or in your Excel 2007 folder

Click the OK button

To use the link, simply click on it

Chapter Eleven   Review

A Hyperlink will let you move quickly to another place inside the document or to a completely different document.

A Hyperlink consists of a link (where you click the mouse) and a destination (where you go after you click the mouse).

Using a defined name for a cell will make it easier to reference the link and the destination. It will also help you remember that the cells are tied together.

A Bookmark is a reference to the destination of the link if it is inside the same document.

1) A Hyperlink will allow you to quickly move to another place in the document, but not to a different document. **True or False**

2) Where, on the Ribbon, will you find the Define Name command? (Give the Tab name and the Group name). _____ Tab _____ Group

3) If you click on a cell but don't know the name of the cell, where can you find it? _____

4) In the Insert Hyperlink Dialog Box, the "Bookmark" button and the "Place in the Document" button both give you the same results. **True or False**

5) On what Tab and Group of the Ribbon is the Hyperlink command located? _____ Tab _____ Group

Chapter Twelve   Workbooks

In this chapter we will discuss the different sheets in a workbook. This will also cover moving between the sheets and renaming the sheets. We will also cover inserting sheets as well as hiding rows, columns, and sheets. We will finish with how to protect your worksheets.

As you have probably noticed every workbook opens with three worksheets, and the average user only uses the first sheet. If information was included on any other sheets, the average user would never know it.

Lesson 12 – 1   Workbook Sheets

Open the Invoice Log.xlsx workbook

This workbook is a listing of all invoices for one Service Company. For this lesson we will separate the invoices by the billing date and put each quarter on a new sheet in the workbook.

Select row 1 on the first sheet

Click the copy button on the Home group of the Ribbon

Click on the Sheet 2 tab at the bottom of the worksheet (See Figure 12-1)

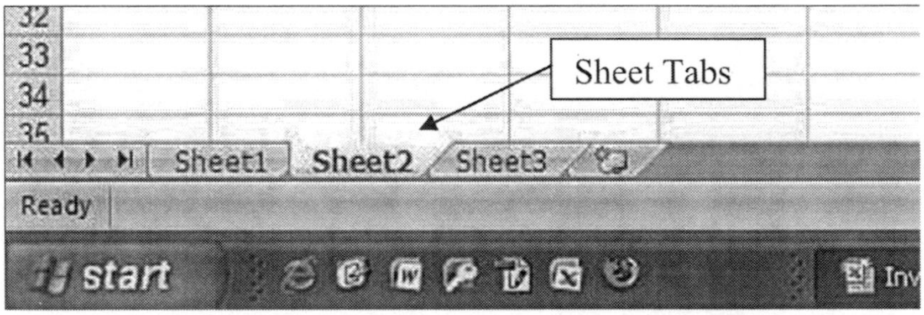

Figure 12-1

Click on cell A1 and then click the Paste button

The main label row will be copied to the second page of the workbook. This is shown in Figure 12-2.

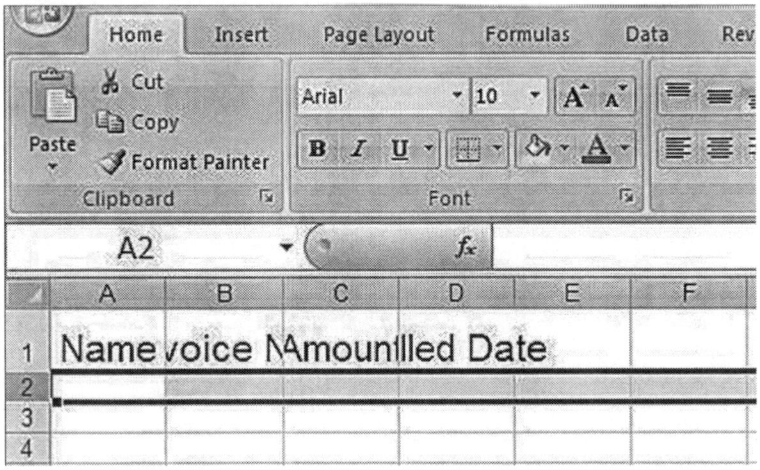

Figure 12-2

This doesn't look quite right, now does it?

Adjust the column widths so they match the following:

        Column A   27.7

        Column B   18.5

        Column C   22.1

        Column D   18.1

Adjust the row heights to that all rows are 26.25

You can adjust the height of all rows at one time by clicking the small square above row 1 and then changing the height of any row. Your worksheet should look like Figure 12-3.

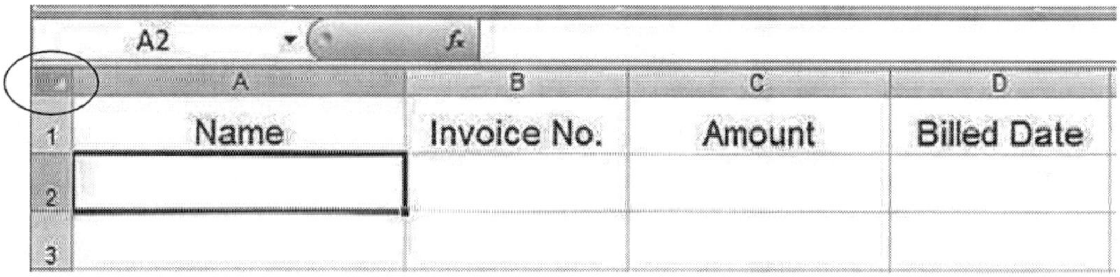

Figure 12-3

Click on the Sheet 1 tab to go back to the first sheet

Select cells A60 through D131

This is done by clicking on cell A60 and holding the left mouse button down and then dragging the mouse until cell D131 is highlighted and then releasing the left mouse button.

Click the Cut button on the Home tab

Click on the Sheet 2 tab, make sure cell A2 is selected, and then click the Paste button

The selected cells are now removed from the first sheet and placed on the second sheet.

You will also have this big blank area in sheet 1. You need to delete this section.

Move back to sheet 1, and while the cells are still selected (even though the cells are empty), click the delete button on the Cells group of the Home tab of the Ribbon (See Figure 12-4).

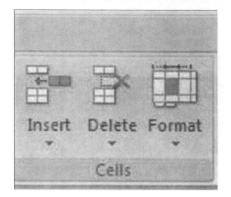

Figure 12-4

This will completely remove the empty cells from the worksheet.

Copy the label cells on the first row to sheet 3 and adjust the columns and rows

Move the cells that represent July, August, and September to sheet 3 and then delete the empty cells from sheet 1

Repeat the above steps to move the cells. The cells start at cell A60 and end with cell D119.

Now we have moved the cells for the second and third quarters, but we still need to move the cells from the fourth quarter to a new sheet. Oh No! We don't have any more sheets to move the cells. Do we have to quit? Only if you really don't like this lesson. Let's add another sheet to the workbook.

Click the Insert Worksheet tab next to the last sheet tab at the bottom (See Figure 12-5)

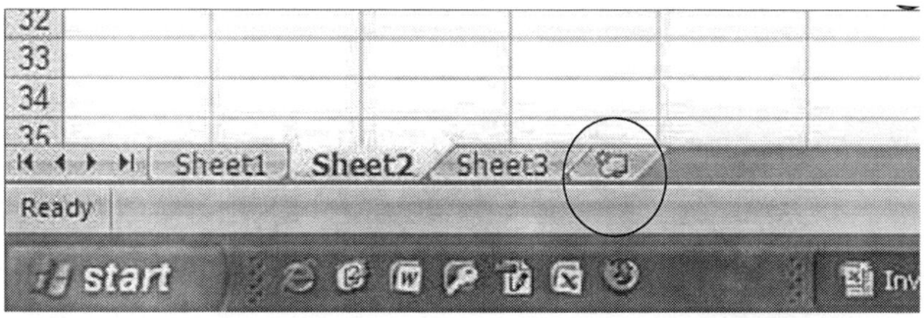

Figure 12-5

Clicking this will add a new worksheet to the workbook (In this case Sheet 4), and bring it to the screen. You will also notice that cell A1 is selected.

Repeat everything you just did to have the October, November, and December invoices on page 4

Okay, we moved all of the cells so that each quarter is on its own page. Now what are we going to do? I must admit that the data like this is not very useful. It would help if each page had a total for the number of invoices and the revenue brought in to the company.

On sheet 1, click on cell A61 (Using this cell will leave a blank line between the cells and the total line) and type Totals

Click on cell B61 and type the formula to count the number of invoices

Okay, I'll tell you how to do it. The formula for counting is as follows:

=COUNT(starting cell name : ending cell name)

Here is the actual formula   =COUNT(B2:B60)

The result in cell B61 should be 58.

This will count the number of cells that have numbers in them, not text.

Click on cell C61 and click the AutoSum button

The AutoSum will try to go to the left because that is the only cell touching it that has a number in it. This is not what we want to add, so we need to tell Excel which cells to add together.

Click on cell C2 and drag down to cell C59 and then click the Enter button

The formula bar should look like Figure 12-6.

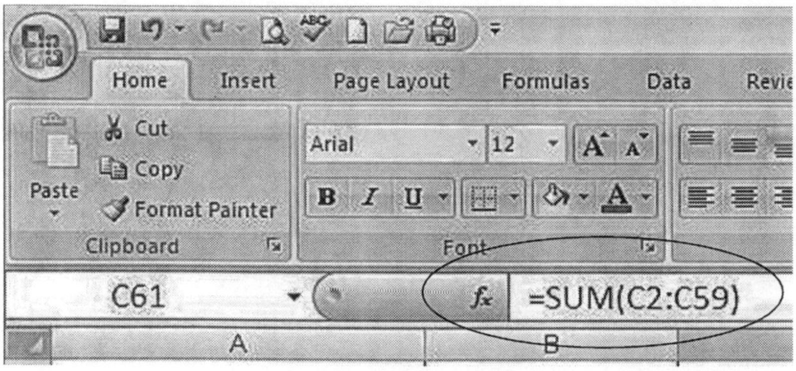

Figure 12-6

Repeat this process to add a totals line to the other worksheets

When you are finished, every sheet should have a total number of invoices for the quarter and the total income for the quarter.

Now we need one more thing we need a sheet to have all of the totals on it.

Add another sheet to the workbook

On this sheet we will put the totals from each quarter and also a grand total for the year.

Fill in the worksheet until it looks like Figure 12-7

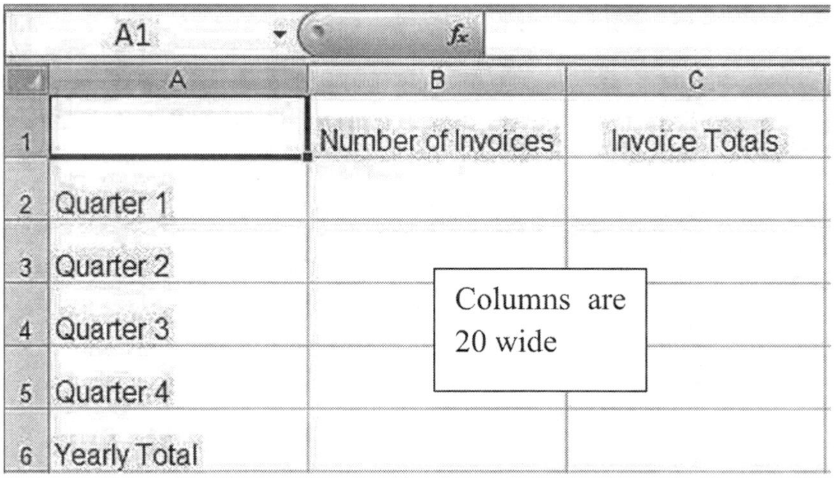

Figure 12-7

Now we need to add the totals. We could just copy the number from the other sheets, but if we made a change in one of the worksheets it would not reflect it on this sheet. We can accomplish what we want by referencing a cell instead of putting in a regular formula.

286

Click on cell B2 of sheet 5

Type the equals sign and then click on the sheet 1 tab and then click on cell B61

Click the Enter button

The Formula bar will look like Figure 12-8.

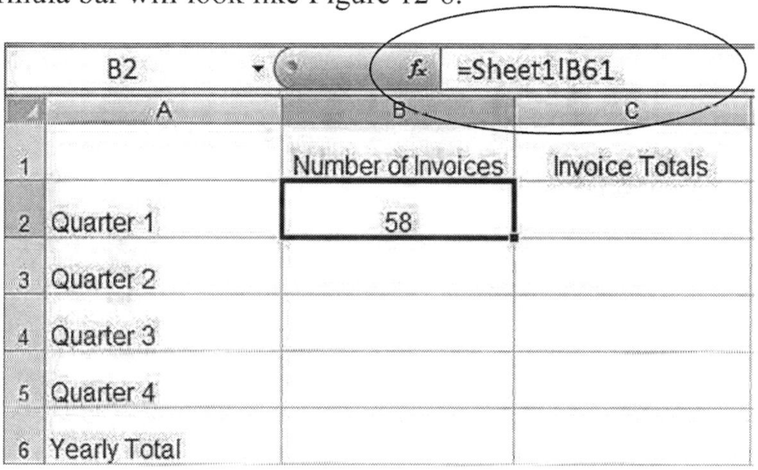

| B2 | ▼ | fx | =Sheet1!B61 |

	A	B	C
1		Number of Invoices	Invoice Totals
2	Quarter 1	58	
3	Quarter 2		
4	Quarter 3		
5	Quarter 4		
6	Yearly Total		

Figure 12-8

You will notice that the sheet number is referenced in the cell. This is because the referenced cell is not on the same sheet as the selected cell. There is also an exclamation point because a cell on a different worksheet is referenced.

Click on cell C2 and reference cell C61 on sheet 1

Repeat this process for the number of invoices and revenue for the remaining quarters

Do an AutoSum for the columns in the Yearly Total cells

Your sheet 5 should look like Figure 12-9

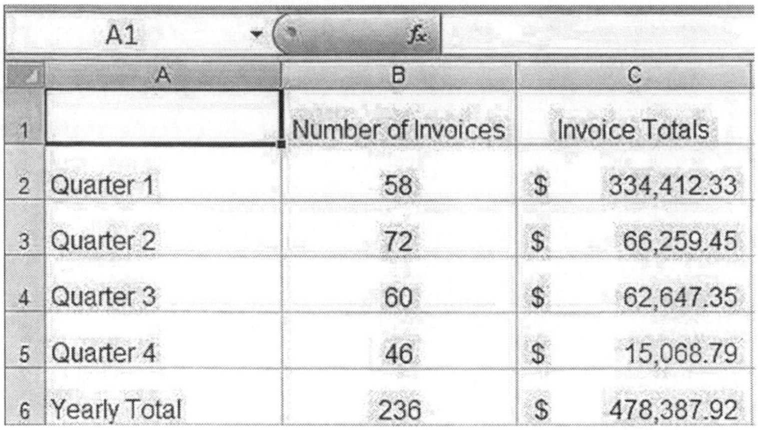

	A	B	C
1		Number of Invoices	Invoice Totals
2	Quarter 1	58	$ 334,412.33
3	Quarter 2	72	$ 66,259.45
4	Quarter 3	60	$ 62,647.35
5	Quarter 4	46	$ 15,068.79
6	Yearly Total	236	$ 478,387.92

Figure 12-9

Note: If you later decide that you don't want a worksheet included in the workbook, it can be deleted by right-clicking on the sheet tab and choosing Delete from the shortcut menu.

Save your work

Lesson 12 – 2   Renaming Worksheets

There are times when it would be easier for the user to understand the different sheets if they had something on them that would easily identify what was on the sheet. Excel will allow us to change the name of the sheet for that purpose. Let's rename the different sheets that we created in our last workbook.

If necessary, open the Invoice Log workbook

It would be a lot easier for the user, if the sheet tabs had a something besides Sheet 1 and 2 on them.

Right-click on the Sheet 1 tab

Select the Rename choice

The shortcut menu is shown in Figure 12-10.

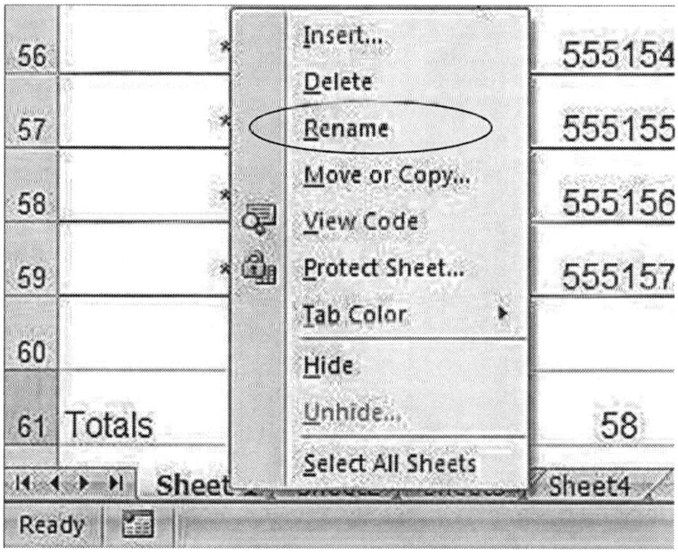

Figure 12-10

After you click the Rename choice, the Sheet 1 name will be highlighted and ready for you to type in a new name. You do not need to click the mouse or anything else, just start typing. This is shown in Figure 12-11

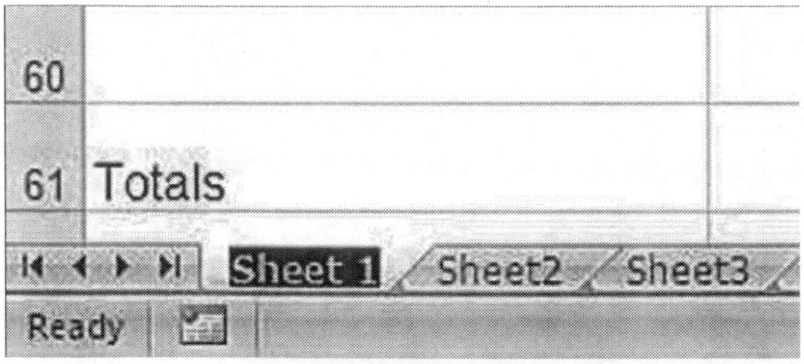

Figure 12-11

Using the keyboard type 1st Qrtr **and press the Enter key**

The name on the tab will change to reflect the name you typed.

Change the names on the other sheets to reflect which quarters they represent and change sheet 5's name to Year Total

Click on the Year Total sheet

Click on cell C2

Notice that the cell reference in the Formula bar does not show Sheet 1 anymore, it shows 1st Qrtr as the sheet to reference.

Excel automatically updated each reference to show the correct sheet name.

Save your changes

Lesson 12 – 3   Hiding Columns, Rows, and Sheets

There may come a time when you would like to not be able to see a column or row or even a sheet. Right now you probably can't think of a good reason to do this, but consider that if you were working with a worksheet with twenty columns and you only needed to look at the first and last column. The other columns need to be there, because they have needed data in them, but you don't have to see them right now. You can hide the columns without disturbing the data in them.

If necessary open the Invoice Log workbook

For this part of the lesson, we will want to hide the amount of the invoice. Hiding a column is relatively easy. You only need to select the column(s) that you want to hide right-click on them and choose Hide from the shortcut menu.

Select column C by clicking on the letter C at the top of the column

Right-click anywhere in the column

Click on Hide from the shortcut menu

The shortcut menu is shown in Figure 12-12.

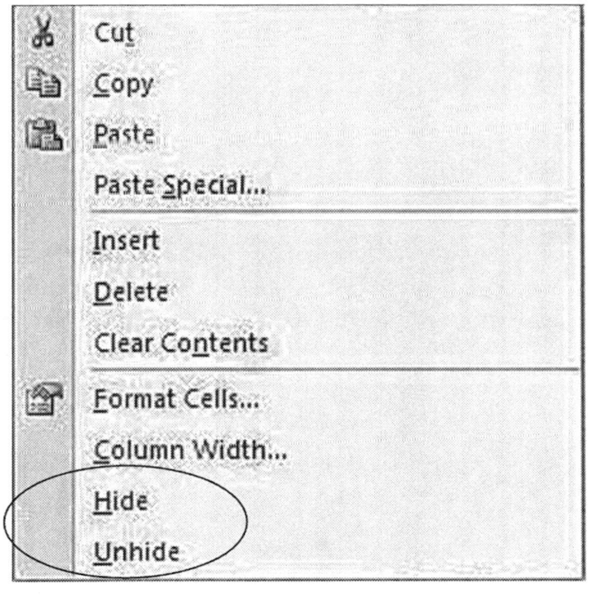

Figure 12-12

You will now see that the entire column is not visible, it is still there but it is not visible. You can tell that it is hidden by looking at the columns. You can see A, B, and D but not C (See Figure 12-13).

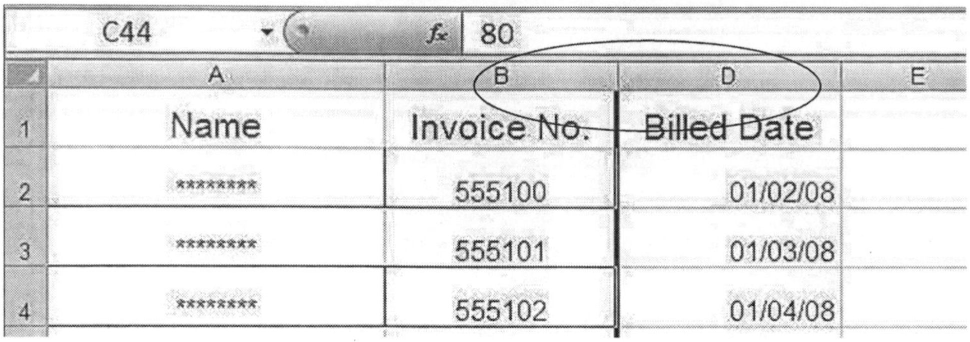

	A	B	D	E
	Name	Invoice No.	Billed Date	
2	********	555100	01/02/08	
3	********	555101	01/03/08	
4	********	555102	01/04/08	

C44 — ƒₓ 80

Figure 12-13

Note: If you want to hide more than one column, click on the first column you wish to hide and drag the mouse to the last column you want to hide. Now you can right-click and choose hide.

Un-hiding a column is just as easy. To Unhide a column click on the column on each side of the hidden column, right-click and choose Unhide.

Click on column B and drag the mouse over to column D

Right-click anywhere inside the highlighted columns and click on Unhide

Everything will go back to normal.

Hide column A

Column A is now gone, but we will need to bring it back at some point. It is going to be difficult to select the columns on both sides of the hidden column this time.

In the Cell Name box type A1 **and then press Enter** (Figure 12-14 shows the Cell Name box)

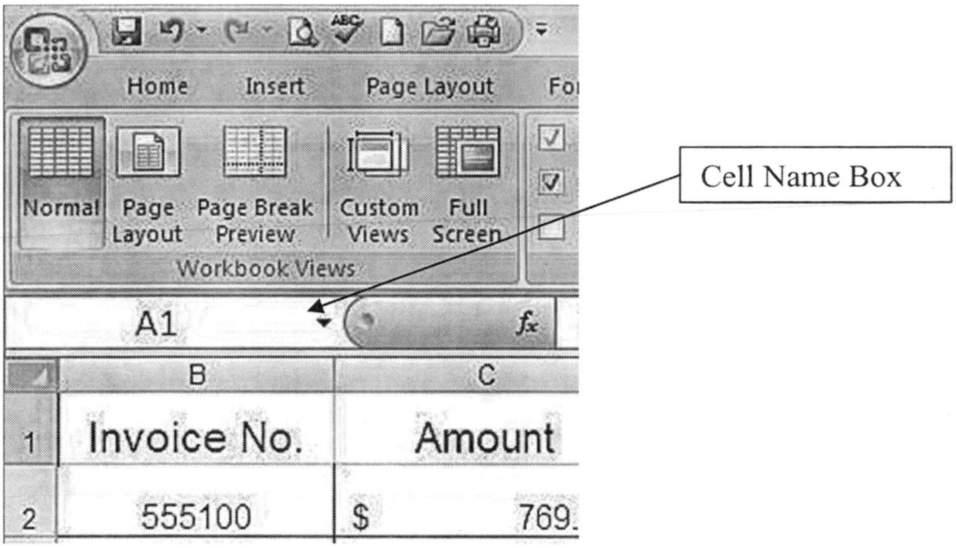

Figure 12-14

The next thing you should notice is that the number 1 to the left of cell B1 it is now highlighted. With this highlighted we can Unhide the A column.

Click the Format button in the Cells group of the Home tab, move the mouse to Hide and Unhide and then choose the Unhide Columns choice (See Figure 12-15)

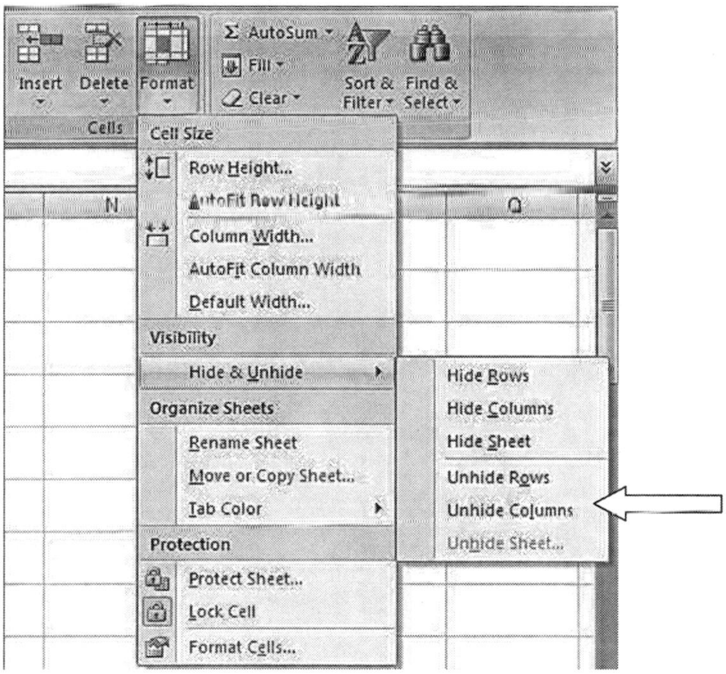

Figure 12-15

293

Column A will immediately return to the screen.

Hiding rows works the same way as hiding columns. You need to select the row(s) to hide, right-click, and then choose Hide from the menu.

Select rows 5 through 8 by clicking on the row number 5 and dragging the mouse to row number 8

Right-click in the highlighted section and choose Hide from the menu

The four rows will be hidden. You can tell by looking at the row numbers (See Figure 12-16).

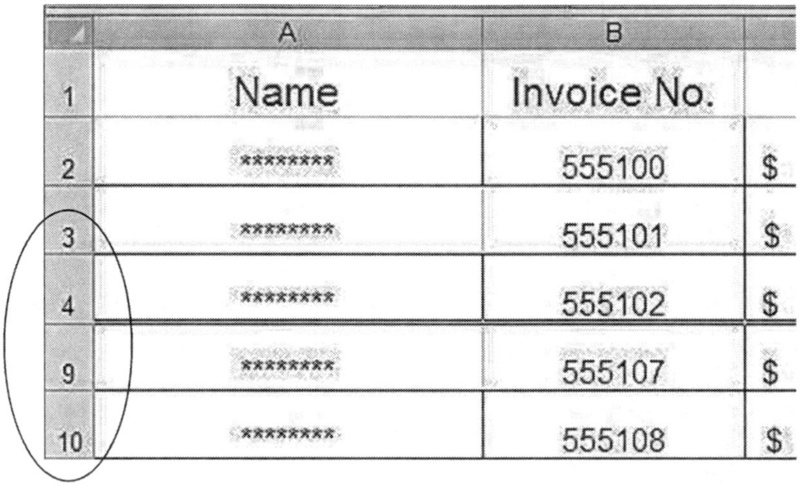

	A	B	
1	Name	Invoice No.	
2	********	555100	$
3	********	555101	$
4	********	555102	$
9	********	555107	$
10	********	555108	$

Figure 12-16

Select rows 4 and 9, right-click and then choose Unhide

The rows magically come back.

If you choose to hide the first row and then want to restore it, you must follow the same process that you used when bringing the first column back into view.

Below are the steps to unhide the first row:

In the Name box type A1 and press Enter
Click the Format button in the cells group
Move the mouse to Hide and Unhide
Select Unhide Rows

Now suppose that we want to hide an entire sheet. For this part of the lesson we do not want anyone to see the Year Total page of our workbook.

Right-click on the Year Total sheet tab at the bottom of the workbook (See Figure 12-17) **and then click on Hide**

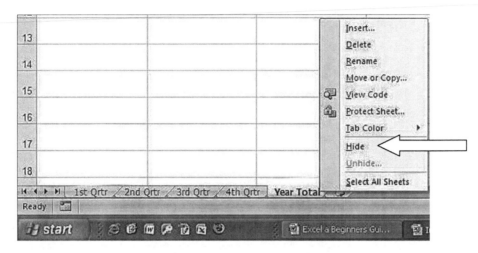

Figure 12-17

The Year Total worksheet appears to have magically vanished. Actually the worksheet is still there it is only hidden.

Now, let me show you how to bring the worksheet back.

Click on the Format button in the Cells group, move the mouse to Hide and Unhide and then choose unhide sheet

This is shown in Figure 12-18.

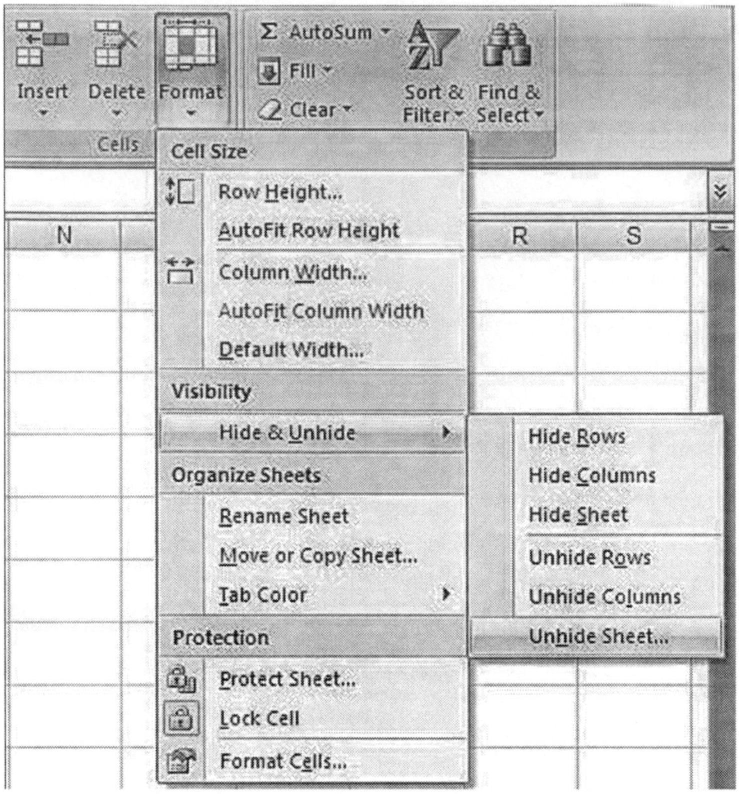

Figure 12-18

Clicking on this will bring a new dialog box to the screen. This dialog box is the Unhide Dialog box and is shown in Figure 12-19.

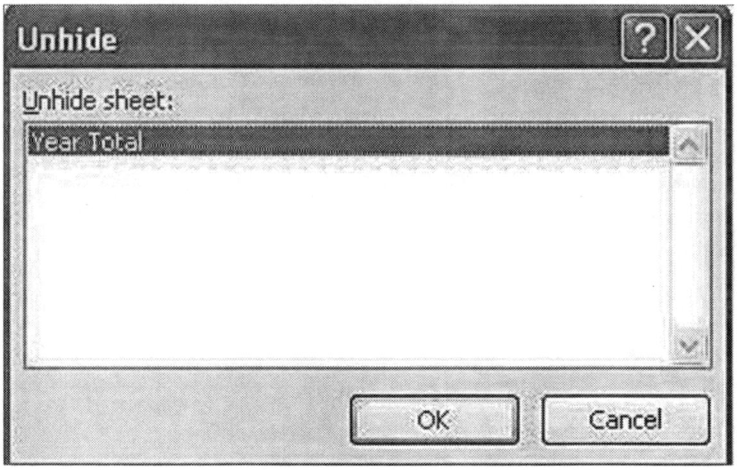

Figure 12-19

In this case there is only one sheet that is hidden, so there is only one sheet listed in the selection inside the dialog box. If more than one sheet was hidden, you would need to click on the sheet you wanted to unhide and then click the OK button. Since the Year Total is already highlighted, all we have to do is click the OK button.

Click the OK button

This will unhide the Year Total worksheet and all is right with the world. If there were no sheets hidden, the unhide sheets choice would not be available.

Save your work

Lesson 12 – 4   Protecting Worksheets

Now comes the time we have all been waiting for. Now we can see how to protect our workbooks. We can keep people from opening the workbook, or making changes to our work. This lesson will show you how.

If necessary open the Invoice Log workbook

Click the Office Button and click Save As from the menu

In the File Name textbox type Protected Workbook **and then click the Save button**

This will allow us to make changes while keeping the original workbook unchanged.

First let's make the entire workbook password protected. This will require the user to enter a password before the workbook can be opened.

Click the Office Button and then move the mouse to Prepare and then click on Encrypt Document

This is shown in Figure 12-20.

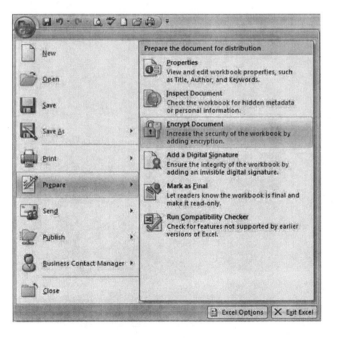

Figure 12-20

You will now have to choose a password that will be required before the workbook can be opened. This is a serious action you are about to make. If you forget this password, you will not be able to open the document. You will also not have any way to retrieve the password. If you are going to use this, write the password down and put it someplace where you can find it when you need it. I would have said to put the password someplace safe, but I never seem to be able to find that safe place a second time, so put it somewhere that you will able to find it. I keep mine in a file folder with Passwords written on the tab. It may not be safely tucked away from the world, but I can always find it.

The Encrypt Document dialog box will come to the screen and is shown in Figure 12-21.

Figure 12-21

Type the word Invoice **in the password textbox and then click OK**

You will notice that the password does not show up as you type it, only a series of asterisks show. This keeps anyone else from seeing what you are typing for a password. Again, notice the warning that it gives you.

Figure 12-22

When you click the OK button, you will have to enter the password a second time to make sure you typed it exactly the way you thought that you typed it. The password is also case sensitive. That means if you used any capital letters, you will have to use the same capital letters every time you open the workbook. You will also have to click the OK button on the Confirm Password dialog box.

Save the changes and close the workbook

Now we are going to open the workbook.

Open the Protected Workbook file

You will notice that the workbook did not open. All you have is a dialog box stating that the workbook is protected and you need to enter the password.

Type the password Invoice **in the text area and press Enter or click OK**

The file will open and you can do whatever is necessary to the workbook.

For the next part of the lesson we will password protect the editing of the workbook.

To make your life easier, we will unprotect the workbook. This will allow anyone to open the workbook, but we will not let them make changes to the last page of the workbook.

Click the Office button, move the mouse to Prepare and then click on Encrypt document

The Encrypt document dialog box will come back to the screen. You will notice that the password is already filled in. You cannot tell what the password is because it only shows as a series of asterisks. If you remove the password so that the text entry is blank, the protection will be removed. Now the workbook can be opened by anyone.

Using the Delete key, remove the password and press Enter

If Year Total is not the active worksheet, click on the Year Total sheet tab

Click the Select All button at the top left side of the workbook

This is the small button above the Row 1 and to the left of column A. This will let us make changes to every cell, row, and column at one time. If you only want to protect some of the cells, you can select them by clicking and dragging the mouse over the cells you want to protect.

Click the Format button in the Cells group of the Home tab of the Ribbon

Click the Format Cells choice at the bottom

This will bring the Custom Lists dialog box to the screen. The last tab in this dialog box is the Protection tab. The choices on this tab will allow us to lock the cells and hide the formulas when we protect this page. The dialog box is shown in Figure 12-23.

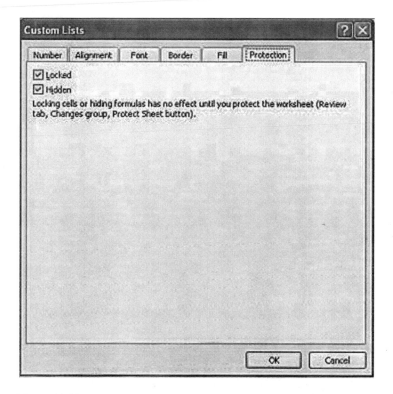

Figure 12-23

For this lesson we will want to lock the cells and hide the formulas. Under the checkboxes is a statement that you will want to read. Nothing you do here will do anything unless you protect the worksheet. You need to understand that this is only the first step in keeping someone from editing your worksheet. In the next step we will finish by protecting the worksheet.

Make sure both checkboxes are checked. If they are not click the small box to the right of each choice.

Click the OK button

Nothing will appear to have happened. When we finish the next step, the changes that we made here will take effect.

Click the Format button in the Cells group again and choose Protect sheet from the menu

This is shown in Figure 12-24.

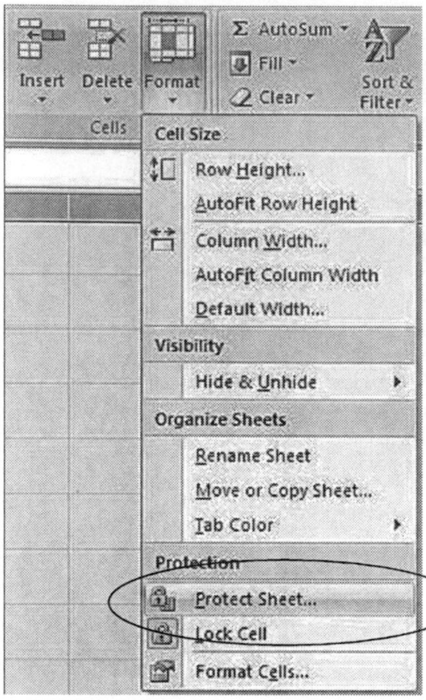

Figure 12-24

When you click this another dialog box will open (See Figure 12-25).

Figure 12-25

This dialog box will allow us to set a password to unprotect the worksheet. We can also choose what actions we want the user to have access to. Some of these choices include can the user select cells, can they format cells, and can the user insert columns and rows?

Type in the password MyInvoice

If any of the checkboxes at the bottom are checked, uncheck them and then click OK

If you leave the first two boxes checked, the user will be able to click on the cell but not make any changes to the data that is there. Also no information will be available in the Formula bar, it will remain blank. If the user clicks inside the Formula bar, an information box will be displayed telling them that the worksheet is protected.

Confirm the password by retyping it in the Conform Password dialog box and then press Enter

The worksheet is now protected and if you try to click on a cell nothing will happen. The cells are now locked and all of the formulas are hidden from the users view. If the user tries to double-click on a cell he/she will get an information box explaining that the cell they are trying to access is protected and they must remove the protection before that can select the cell

The process for removing the protection is similar to adding the protection.

Click the Format button again and this time choose Unprotect sheet

The dialog box to enter the unprotect password opens. All we have to do is enter the correct password and the sheet will no longer be protected.

Type the unprotect password MyInvoice and press Enter

Now the worksheet is back to normal.

The protection will not come back automatically if you leave the worksheet. Also if you save the changes without re-protecting the worksheet, it will not be protected the next time someone opens it.

Note: You can encrypt the workbook so that you need a password to open it, and you can also protect the sheet so that you need a password to edit the cells. These two things are totally separate and you can use one or both in one workbook.

Protect all of the sheets so no one can make changes to the workbook and then encrypt the entire workbook

Make sure you record the passwords somewhere, so you can open the workbook and make changes if you want to. If you want each page can have a different password.

Save your changes

Chapter Twelve   Review

Worksheets are the individual sheets inside the workbook. When you open a workbook there will usually be three worksheets showing.

New information can be entered into a worksheet or data can be copied from an existing worksheet.

New worksheets can be added by clicking the Insert Worksheet tab.

Cells on one worksheet can reference cells on a different worksheet. Formulas can also reference cells on a different worksheet.

Worksheets can be renamed to give the user an idea as to the worksheets contents.

Columns, Rows, and Worksheets can be hidden if you do not want them to be seen, and unhidden later if you need them to be seen.

Workbooks can be password protected to keep anyone from opening the workbook or to keep anyone from making changes to the worksheet.

Chapter Twelve   Quiz

1) To insert a new worksheet you need to click on the Worksheet command that is located on the Insert Tab of the Ribbon. **True or False**

2) Adjusting the column width on sheet 1 will automatically adjust the width for that column on all worksheets. **True or False**

3) You have selected several cells on a worksheet, How do you remove the cells from the worksheet?

4) Write the formula to count the number of cells with numbers in them from cell C4 to cell C82.

5) Write the formula for a cell in worksheet 1 to equal the same value that is in cell E6 of worksheet 2.

6) Worksheets can be renamed by clicking on the Rename command on the Page Layout Tab of the Ribbon. **True or False**

7) After hiding a column or row, it is impossible to tell that it is hidden. **True or False**

8) To unhide column A, what is the first thing you must do?

9) If you password protect your workbook and you forget or lose the password, it cannot be retrieved. **True or False**

10) To make changes to a worksheet that has been password protected you must _____ the worksheet by providing the correct password.

Chapter Thirteen   More on Printing

This may sound like a useless chapter. After all what is the big deal? You either print or you don't, right? In this chapter we will go beyond just clicking the print icon on the Quick Start Toolbar. You will learn how to use the Page Setup Group, which is located on the Page Layout tab of the Ribbon. This contains the items which were previously available when you clicked the File button and then clicked the Page Setup item from the Menu bar of previous versions of Excel.

Lesson 13 – 1   Margins

The Page Setup Group is located on the Page Layout tab of the Ribbon and is shown in Figure 13-1.

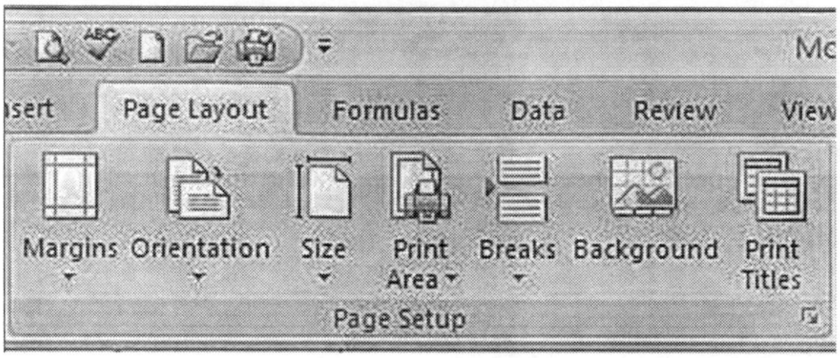

Figure 13-1

I waited until this chapter to use the Page Setup group because these are normally the things you will want to use before you print a document.

The first thing we will want to do is check the margins. The margins are the distance between the sides of the page and the edge of the worksheet. We will check these to see if they might need to be changed so the workbook will print on one page. If the workbook will not quite fit on one page when it is printed, we can change the margins just a little and it might fit.

Open the Page Setup workbook

This can be found in the downloaded files or in the Excel 2007 folder, if you copied the files to this folder.

Before we start making any changes to anything, it would be nice to see if the width of the worksheet will fit on a single when printed.

Click on the Print Preview Command

This command is located by clicking the Office button and moving the mouse down to print and then clicking on Print Preview (See Figure 13-2).

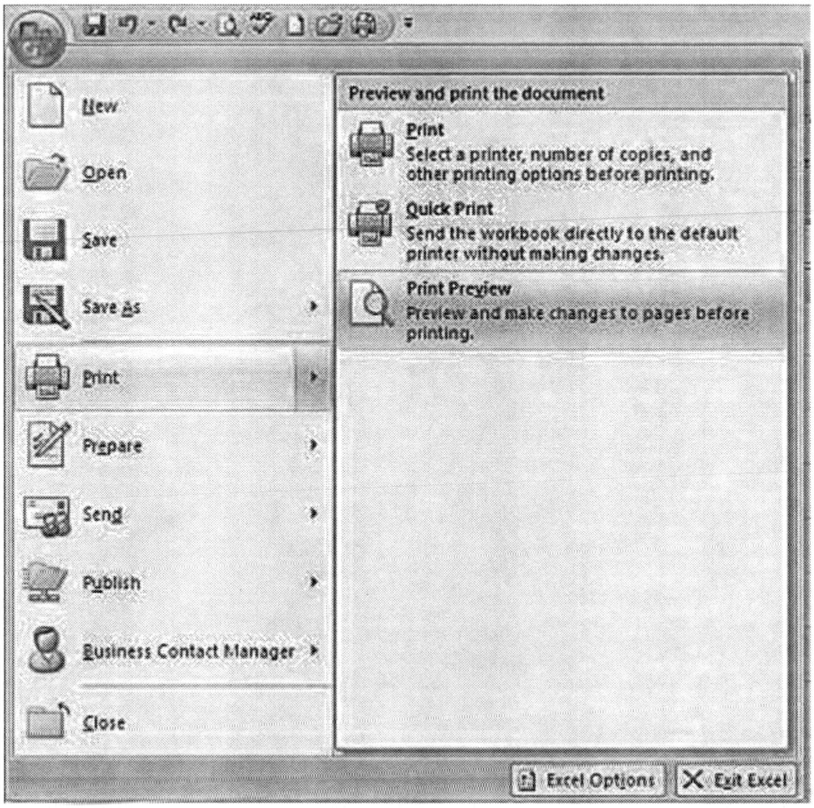

Figure 13-2

Selecting Print Preview will allow us to see what the worksheet will look like when it is printed. We will be able to see if changing the margins will have any effect on the printed sheet.

As you can see, the first sheet will only be able to print up to the year 2005, 2006 will not be on this page.

Click on the button that says Close Print Preview by the top

The Print Preview screen will close and you will notice that there is a dotted line between columns H and I. This dotted line will show us exactly where the printable area stops. We can see if adjusting the side margins will allow the entire width of the worksheet to be displayed.

Click on the Margins button of the Page Setup Group of the Page Layout tab

This is shown in Figure 13-3.

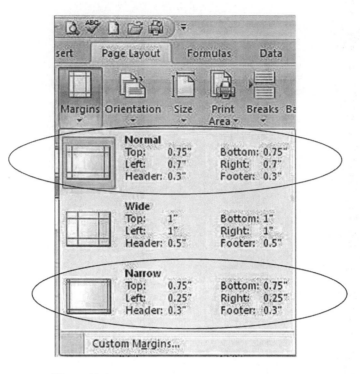

Figure 13-3

You will see the differences between the Normal margins versus the Wide and Narrow margins.

Click anywhere in the Narrow section

As you move the mouse over the choices the entire section will become highlighted and then you can click the mouse anywhere inside the highlighted area to use that selection.

The dotted line will shift to the right and the area that will show on the printed page will now show column I on the page.

There may be times when these choices are not sufficient for your needs. There may be times when you want to adjust the top margins and not just the sides. The Custom Margins choice will address this.

Change the margins back to the Normal settings

Click the Margins button again and this time click Custom Margins at the bottom of the menu

This time the Page Setup Dialog box will jump onto the screen as shown in Figure 13-4.

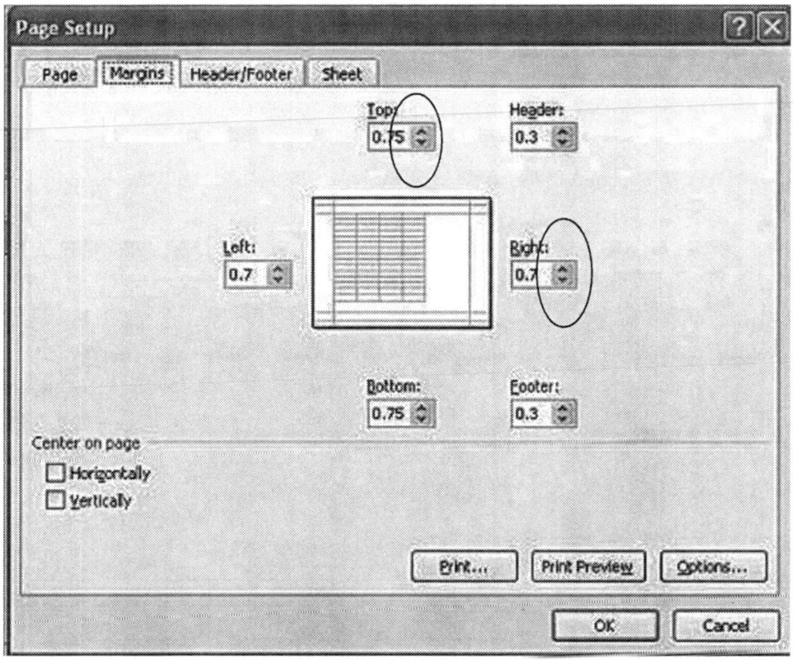

Figure 13-4

The Margins tab will be selected and you can now manually change any of the margins.

Using the up and down adjustment arrows (circled in the figure) change the left and right margins to 0.25 and the top and bottom margins to 0.5

From this dialog box you can also change the Header and Footer margins. You will notice that they are above and below the top and bottom margins. You can also center the worksheet vertically and/or horizontally on the page.

Click OK when you have finished

Close the Workbook

Lesson 13 – 2   Orientation

This lesson will deal with how the page is printed on the paper and what size paper is being used in the printer.

If necessary open the Orientation workbook

This workbook is similar to the Monthly Budget workbook. The column widths have been changed slightly to accommodate this lesson.

Click the Print Preview choice from the Office button and Print choice

You will immediately see that the entire worksheet will not fit on one page.

Click the Next Page choice by the top (See Figure 13-5)

Figure 13-5

The part of the worksheet that did not fit on the first page will be printed on the second page. As you can see, simply changing the margins will not fix this problem. We really need this to be printed on one page so the user will be able to see all of the information at one time.

To make this happen, we can change the orientation of the page. Normally when you put something on an 8 ½ x 11 sheet of paper, the 8 ½ is the width and the 11 is the length. When we print like this it is called Portrait orientation. If we reverse these it is called Landscape orientation.

Click the Close Print Preview button

You will now be able to see the dotted lines identifying the width of each page. If we change this worksheet to a landscape orientation, we will be able to print the entire page on one sheet.

Click the Orientation button in the Page Setup group (See Figure 13-6)

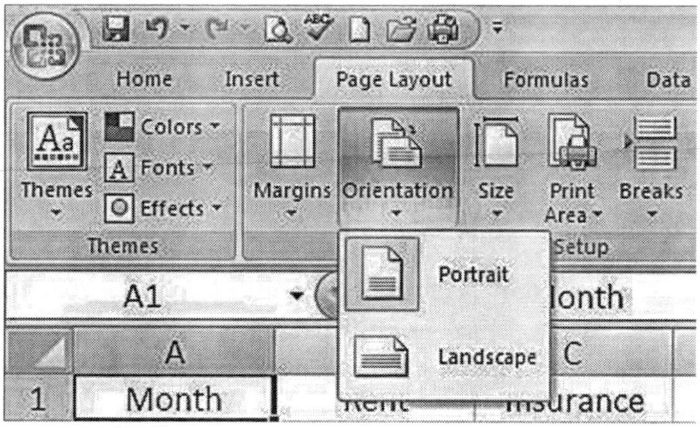

Figure 13-6

The default choice is Portrait, but that will not give us what we want. To get the worksheet on one page we will need to choose Landscape.

Click on the Landscape choice

The dotted line will move to the end of the food chain (sorry food column). Now when we do a print preview the whole sheet will show on one page.

Click the Print Preview menu choice again

You have verified that this will indeed print on one page.

Click on Close Print Preview

Save and close the Orientation workbook

Lesson 13 – 3   Paper Size

In this lesson we will face a different problem, what if the width is not the problem? What if the worksheet is just too long for the page?

Open the Legal Size workbook

This lesson will show you how to change the settings to print to a larger size of paper (legal size) than the standard letter size. I probably don't need to say this but I will anyway. To actually print to legal size paper, you need to change the actual paper in the printer from letter size to legal size.

Do a print preview so you can see the dotted lines showing where the pages end

As you can see, the worksheet is longer than the standard letter size paper. If you try changing the orientation to landscape, it will not help. It will still take two pages to print the worksheet. What we need is a paper that is a little longer that the standard paper.

Click on the Size button in the Page Setup group (See Figure 13-7)

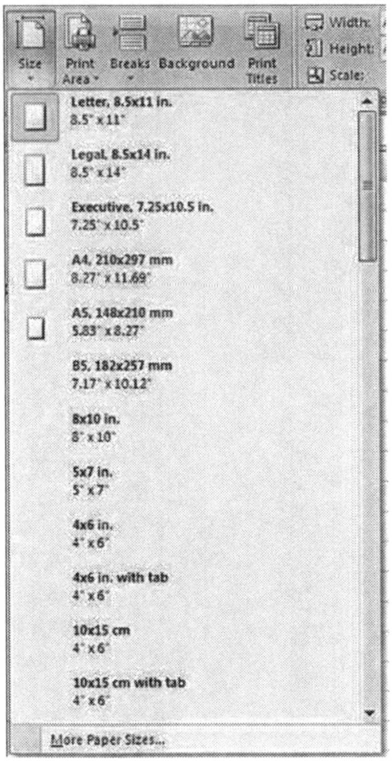

Figure 13-7

The default value is Letter size and it is highlighted. We need something a little longer, so we will choose legal size.

Click on Legal Size

The computer will now make an assumption. It will assume that you actually changed the paper to legal size. The computer changed its internal settings so that it will be able to print to the longer legal size paper.

Use the scroll bar on the far right side of the screen to scroll down and see that the worksheet will fit on one legal size sheet

Note: If you did not change the paper, only one sheet will be printed not two. Also the printer will print to the very end of the sheet. It will not stop at the regular margin.

Save your changes and close the workbook

When you save the changes, all of the changes you have made will be saved. This includes the size of the paper the computer expects to be in the printer when it prints this worksheet. This will not change the default paper size for new workbooks, they will remain letter size.

<h1 style="text-align:center">Lesson 13 – 4   Setting Page Breaks</h1>

Sometimes things just don't seem to fall exactly where you want them. You may have a worksheet that no matter what you do it won't print on one sheet. And it always seems that the page break is never in a convenient place. In this lesson you will learn how to insert the page break where you want it to fall.

Open the Page Break workbook

Do a print preview so you can see where the page breaks will fall and then close the print preview page

As you saw, the page breaks are not in a convenient place. It would be better if they were just under the monthly average row, not in the middle of the month.

When you insert a page break, it will be immediately above the row of the cell that is selected. It will be easier to understand once you perform the action, so let's insert a page break.

Click on cell A39

Click on the Breaks command in the Page Setup group (See Figure 13-8)

Figure 13-8

Click on Insert Page Break

The page break will be immediately above row 39 (remember we selected cell A39). If you ever need to, you can remove a page break the same way you inserted one.

Save your changes

Lesson 13 – 5   Printing Titles

This lesson may seem a bit strange. Why would you want to print titles? In this lesson you will learn how to print the headings for the rows on each worksheet.

If necessary open the Page Break workbook

Do a print preview to see what the printed page will look like

Look at the difference between pages one and two. Page one has the heading at the top identifying what is in each column. Page two does not have any headings at the top. It would certainly look better if there was a heading at the top of the second page. Let's add one.

Click on Print Titles in the Page Setup group

The Page Setup dialog box will come to the screen as shown in Figure 13-9.

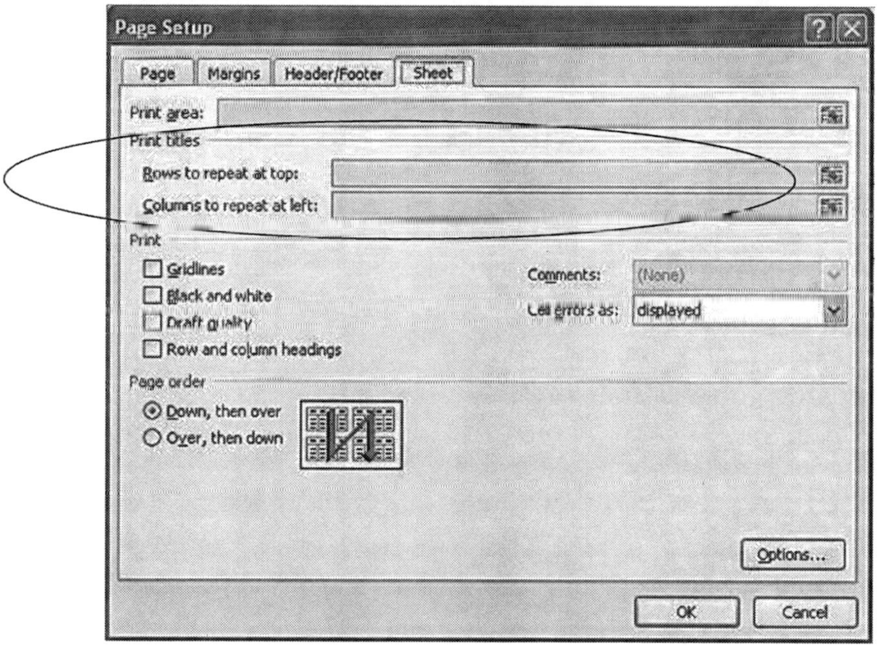

Figure 13-9

This time the Sheet tab is at the top. From here you can make several choices. The one we are interested in right now are the Rows to repeat at the top and the Columns to repeat at the left. Since Excel does not know whether you put the heading that you want on each page at the top or down along the left side, you get to make the decision. In our case we want to repeat what is in the columns at the top. That means that we will have to repeat the first row on each page. Actually it will look better if we also repeat the blank row between the headings and the data, so we will repeat the first two rows.

Click inside the text area of Rows to repeat at the top

Now we can select the rows that we want to repeat at the top of each sheet.

Click on the number 1 on the left side and hold down the left mouse button while you drag the mouse down over the number 2 and then release the left mouse button

When you release the left mouse button you will have what is shown in Figure 13-10.

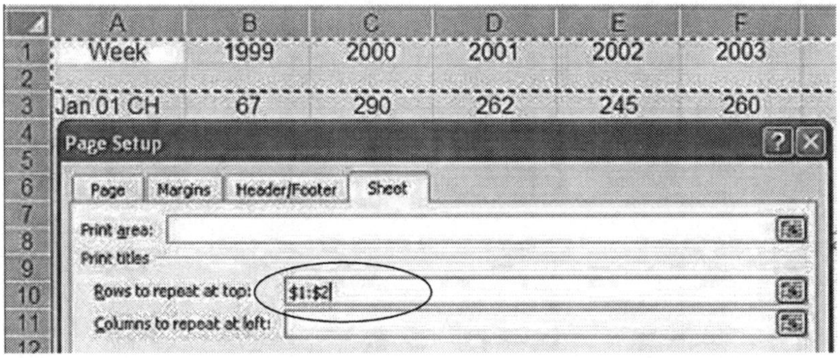

Figure 13-10

Excel put the reference to the first two rows into the text area and now these two rows will be at the top of each printed page as soon as you click OK.

Click the OK button

Do another print preview and check out the results

There are a few other things that you can do from the Sheet tab of the Page Setup dialog box. I will go over these somewhat quickly. In the Print section you can choose to print all of the gridlines so that each cell will have a border around it. Also if you click the checkbox next to Row and Column headings, the printer will print the A B C etc. across the top and the 1 2 3 etc. down the left side. You probably won't want these printed.

Save your changes

Chapter Thirteen   Review

You can use the Print Preview command to see how the worksheet will layout on the printed page. Margins can be adjusted, if required, to have the worksheet fit on a single page. Use Custom Margins if the preset margins will not bring the desired results, or if more exact adjustments are needed.

You can use the Orientation command to change how the worksheet is printed on paper. Portrait has the long side of the paper going up and down while Landscape has the long side of the paper on the top and the bottom.

If the default paper size is not long enough for the worksheet to be printed on one sheet of paper, you might want to consider changing the paper and setting to legal size.

Title can be printed across the top of the pages or down the left side if needed.

Chapter Thirteen   Quiz

1) Margins can be changed if necessary. This is done from the Data Tab on the Ribbon. **True or False**

2) After you close the Print Preview screen, the printable area is shown by a dotted line. **True or False**

3) When you change the margins on the paper, you only have three choices: Normal, Wide, and Narrow. **True or False**

4) Name the two options for page orientation. _____ and _____

5) There are only two preset paper sixes: Letter and Legal. **True or False**

6) Where on the Ribbon do you find the Page Break command? _____ Tab _____ Group

7) Once a Page Break is set, it is there permanently. **True or False**

8) Name an advantage to having Titles printed on the worksheet.

Chapter Fourteen　　　　Macros

If you find yourself performing the same keystrokes over and over you may want to record a macro that will perform these keystrokes for you, then you can run the macro whenever you need to perform these keystrokes. I should inform you that clicking on the Ribbon will not be included in the macro. The Ribbon is a component of the Microsoft Office Fluent user interface and using it cannot be a part of the macro.

Lesson 14 – 1   Recording a Macro

Open the Macros workbook that is part of the downloaded files

This workbook uses the Average formula several times. We will create a macro to type the Average formula in the cell for us.

Click on cell B8

This is the first cell that will have the average formula in it. Now we must tell Excel that we want to create a macro.

Click on the View tab of the Ribbon

Click the down arrow on the Macro button (See Figure 14-1)

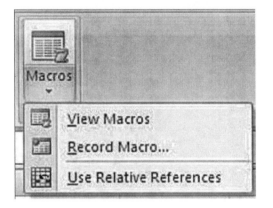

Figure 14-1

You create a macro by recording it. There are a couple of things you need to do before you actually record the macro. These include giving it a unique name and if you want, give it a shortcut key.

Click on the Record Macro choice

The Record Macro Dialog box will pop onto the screen and is shown in Figure 14-2.

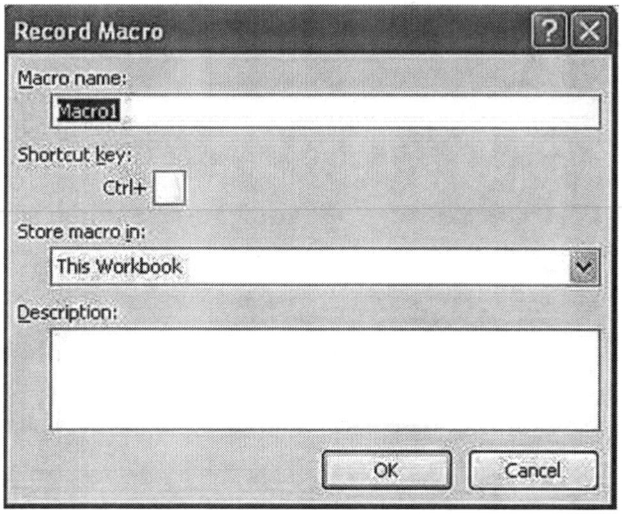

Figure 14-2

The first thing you do is give the macro a name.

Using the keyboard type AVG_Macro **but do not press the enter key**

We need to finish filling out the dialog box before we press the Enter key. The next thing we will want to do is create a shortcut key to run the macro. If we don't create a shortcut key there is about four steps to run the macro and it is just about as easy to type the command as do four steps.

Click in the box beside Ctrl + and then type the letter y

Click in the Description textbox and type

This will input the formula to find the monthly average.

Before you click the OK button you need to understand that every keystroke and every mouse movement will be recorded until you stop recording.

Click the OK button and then follow all of the following steps exactly

Type the equals sign

Press the A and then the V key

Double-click on the word AVERAGE **that is just below the cell**

Click on cell B3

Click on cell B3 a second time and drag the mouse down to cell B7

Click the Enter button

Click on the Macro button arrow and choose Stop Recording

The macro is now recorded and ready to use.

Here is what we did. First, all formulas must begin with an equals sign. As we typed AV, the pre-made formula choices were brought forward so we could choose one. We double-clicked on AVERAGE to enter the word into the formula. We clicked on the first cell in the range we needed to include in the formula. Then we drug the mouse down to the last cell in the range. We then entered the formula into the cell by clicking the enter button.

We gave the macro a shortcut key(s). These two keys when pressed at the same time will run the macro (actually you hold the Ctrl key down and then press the second key).

Click on cell C8 and then hold the Ctrl key while you press the y key and then release both keys

Cell C8 should now have 239 in it and the formula bar should have =AVERAGE(C3:C7) in it. Any time you want to use the macro all you will have to do is select the cell and use the shortcut keys.

Save your changes

Lesson 14 -2   Adding a Macro to the Toolbar

There is an easier way to run a macro that using the shortcut key. You can add the macro directly to the Quick Start Toolbar. This lesson will show you how to add the previously recorded macro the toolbar.

If necessary open the Macros workbook

Click the small down arrow at the right end of the Quick Access Toolbar (See Figure 14-3)

Figure 14-3

This will bring the Customize Quick Access Toolbar drop down menu to the screen as we discussed in Lesson 1 – 5.

Click the More Commands choice toward the bottom

This will bring the Excel Options Dialog box to the screen.

Click the down arrow under Choose Commands From (the default choice is Popular Commands) and then click on Macros (See Figure 14-4)

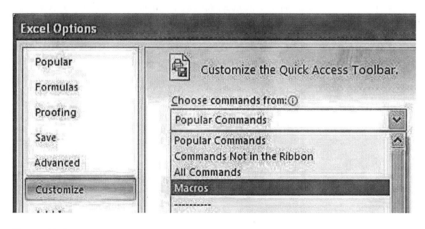

Figure 14-4

This will bring all of the recorded macros to the screen.

Click on the AVG_Macro and then click the Add button in the center

Click the OK button

The Quick Access Toolbar now has a new member and it should appear just to the left of the down arrow. This is shown in Figure 14-5.

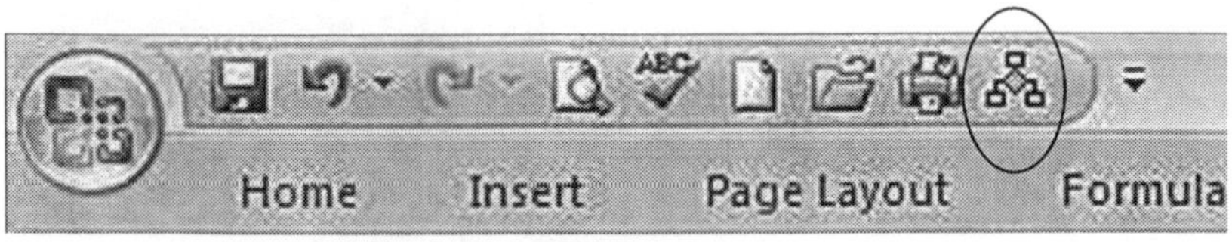

Figure 14-5

The toolbar in the figure may not appear exactly like the one on your computer. I could have different commands on my toolbar, so it may appear slightly different.

Let's test it and see if it works.

Click on cell D8

Click on the macro button on the toolbar

Cell D8 should now show the average for the month and the formula should be reflected on the Formula Bar.

Save your changes and close the workbook

Chapter Fourteen   Review

Using a macro will keep you from typing the same keystrokes over and over.

Once you give the macro a name and click the OK button, or press Enter, every keystroke and mouse movement will be recorded until you stop recording. This, however, does not apply to accessing the Ribbon.

Using the shortcut keys will let you run the macro by pressing two keys on the keyboard.

You can add a shortcut to the Quick Access Toolbar to run the macro with only a mouse click.

Chapter fourteen   Quiz

1) All keystrokes and mouse movements, even accessing commands on the Ribbon, are recorded when you are recording a macro. **True or False**
2) If the recorded macro no longer suites your needs, you can edit the macro by using the Edit command in the Macro Group. **True or False**
3) The macro shortcut keys work with the ALT key on the keyboard. **True or False**
4) Adding a macro to the Quick Access Toolbar is common, so it is found under Popular Commands. **True or False**

Chapter Fifteen   Working with Views

Working with view is not something the average user bothers to try. Some who try do not see any value in using this feature, so they never try it a second time. I hope by the end of this chapter your view changes.

Lesson 15 – 1   The Different Views

Most of the time you will use the Normal view when working with Excel documents. There will come a time when you realize that another view will make your life a little easier. In this lesson you will get to try the different views that are available.

Open the Views workbook

This workbook is found in the downloaded files. We will be able to use this to try out the different views that are available to us.

Click on the Page Layout View

The different views are found on the View tab of the Ribbon and are located in the Workbook Views group (See Figure 15-1).

Figure 15-1

This view will let you make last minute adjustments to your worksheet before you print it. With this view you can still change the formatting and the general layout, but this view also lets you use the ruler to measure the height and width of the columns and rows. The Ribbon and all of its contents are still available for you to use. This will give you a chance to see how it will look when it is printed and still have all of the features of the Ribbon at your disposal.

When you are finished all you have to do is click back on Normal view.

Click on Normal View

If you want to see how the page breaks fall within your workbook, you can use the Page Break View.

Click on the Page Break View

With this view, you can see exactly where the breaks will be on the worksheet when it is printed. You will also be able to see the order the sheets will be printed. You will also have a dialog box that explains how to adjust the page breaks if you need to adjust them (See Figure 15-2).

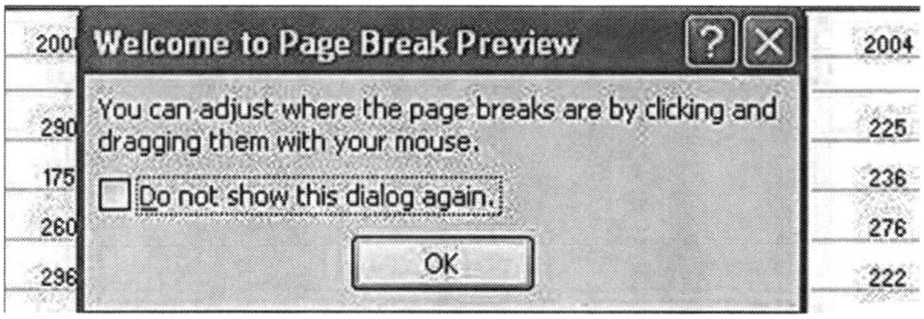

200	**Welcome to Page Break Preview** [?][X]	2004
290	You can adjust where the page breaks are by clicking and dragging them with your mouse.	225
175	[ ] Do not show this dialog again.	236
260		276
296	[OK]	222

Figure 15-2

Click back on Normal View

The time will come when you wish that you had more room on the screen to work on the worksheet. The Full Screen View might be just what you are looking for.

Click on the Full Screen View

Wow I bet you didn't expect that now did you? The Ribbon is gone, the Toolbar is gone, the Office button is gone, and the Status bar is gone. Excel is giving you as much room on the screen as possible to work on the worksheet. Now for the bad news, all of the things that are gone are gone. They are not there for you to use in this view.

When you are finished using this view, you cannot click on one of the other views. That makes sense since they are not there for you to click on them.

Right-click anywhere inside the worksheet

When you right-click inside the sheet a menu will appear with all of the commands that are available to you (See Figure 15-3).

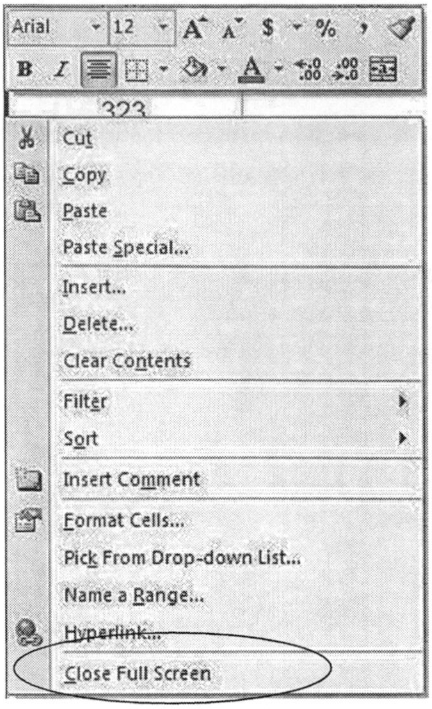

Figure 15-3

To return to the Normal screen, click on the Close Full Screen choice from the menu.

Click on Close Full Screen

You will go back to the view you had before you went to the Full Screen View.

Next we will mention the custom views button.

Custom views are different than you might expect. You don't get to design a new view from a list of possible things you would like to see. Instead it allows you to save the current view under a new name and then apply it to any worksheet that you want.

Suppose you wanted a way to quickly hide three columns in the worksheet. Let's hide the three columns and then create a view to quickly change the way the worksheet is displayed on the screen.

Hide columns D E and F

If you do not remember how to do this refer back to lesson 12-3.

Click on the Custom Views button

The Custom Views Dialog box will come to the screen. It is shown in Figure 15-4.

Figure 15-4

You can add the current view to the list of Custom Views by clicking the Add button.

Click the add button

The Add View Dialog box will come on the screen as shown in Figure 15-5.

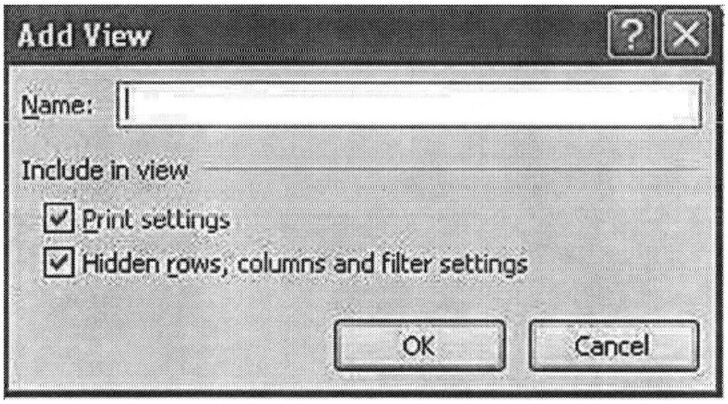

Figure 15-5

This is the place where we give our view a name. We need to have a unique name. Let's call this view the Hidden view.

Using the keyboard type Hidden **in the area provided**

Click the OK button

Now we can try it and see if it works.

Unhide columns D, E, and F

333

With everything back to normal, we can see if this custom view stuff really works.

Click on cell A1 (We don't want anyone to think we cheated by doing something sneaky while the cells were highlighted).

Click Custom Views and make sure Hidden is highlighted and then click Show

Cells D, E, and F are now hidden. Boy that was easy and just think anytime you want to hide columns D, E, and F, you can do it with the click of the mouse.

Note: If you unchecked the "Include hidden columns and rows" button, the columns would not have been hidden.

Unhide the columns

Save your changes

We don't want the columns to remain hidden in the workbook, but we do want the view to be saved in the workbook.

Lesson 15 – 2   Show and Hide

The Show and Hide Group of the View tab will determine what we see on the screen while we are working on the worksheet. We will explore these things in this lesson.

If necessary open the Views workbook and then click on the View tab

The Show and Hide group is shown if Figure 15-6.

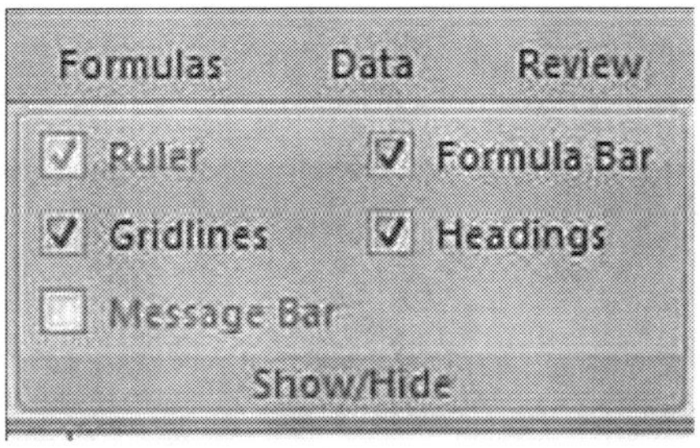

Figure 15-6

These checkboxes tell us if something is visible. If the box is checked, the item is visible.

Click the Gridlines checkbox

All of the Gridlines around the cells are now invisible.

Click the Gridlines checkbox again

All of the Gridlines are back.

Click the Formula Bar checkbox and then click it a second time

These are toggle switches. If the item is checked when you click it, it changes to unchecked and vice-versa.

Click the Headings checkbox

All of the headings are now gone, both for the columns and the rows.

Click the Headings checkbox a second time

Now they are back. It is almost like magic.

Lesson 15 – 3   Using Zoom

If you have ever had someone make an Excel spreadsheet and the cells were so small that you had trouble seeing them, you will like the zoom feature. Perhaps it is just my eyes getting bad, but I like the Zoom feature.

Open the Views workbook if it is not open

In Excel there is more than one way to do something, so we will do it the hard way first. There is a Zoom group on the View tab and it is shown in Figure 15-7.

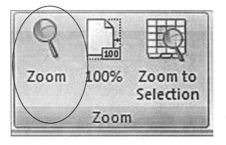

Figure 15-7

Click on the first Zoom button

The Zoom Dialog box will come to the screen and is shown in Figure 15-8.

Figure 15-8

From here you can decide just how much magnification you want to apply to the view. By default the zoom is set to 100%, but this can be raised or lowered.

Click on 200% and press OK

Now do it again and choose 50%

Now click on the center button that says 100%

Everything is back to normal.

Select cell A1 and then click on Zoom to Section

This will bring the Zoom to its highest zoom and focus on the selected cell.

Click the 100% again

I said there was a hard way, so there must be an easy way. At the far right side of the Status bar is a Zoom control (see Figure 15-9).

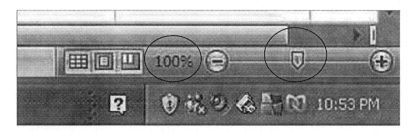

Figure 15-9

Click on the slider bar in the center of the control

Move the slider toward the + and then toward the –

You can watch as you zoom in and out on the page. If you want to know what the percentage is, watch the number on the left.

Do not save any of the changes you have made

Lesson 15 – 4   Freezing Panes

The Freeze Panes command is located in the Windows group of the View tab. Freezing a pane will allow it to remain on the screen while other rows or columns are allowed to scroll out of site.

Open the Views workbook and unhide any columns that might be hidden

Click on the View tab

Click on the Freeze Pane button

A drop down menu will appear and you can choose what you want to freeze. Since they all work the same, we will see how to freeze a column.

Click on the Freeze First Column choice at the bottom (See Figure 15-10)

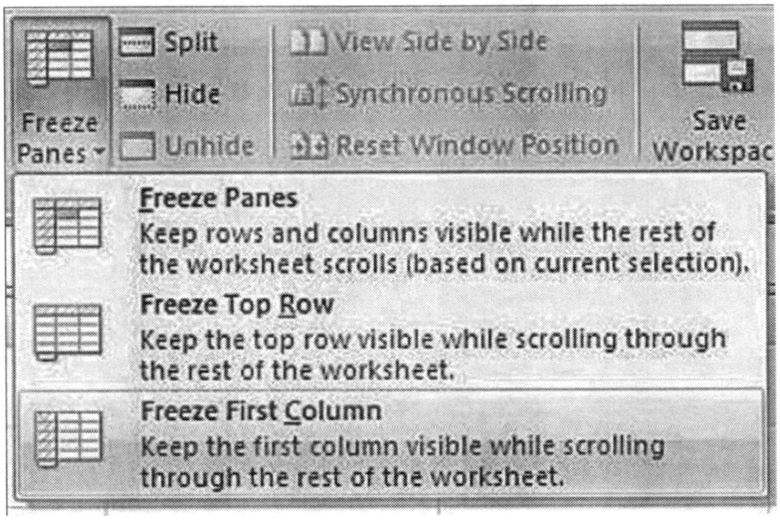

Figure 15-10

Column A now has a line on the right side of it. This is where the freeze starts.

Using the scroll bar at the bottom of the page, just above the Zoom control, move the columns to the left by clicking on the small arrow on the far right side.

See Figure 15-11 for details.

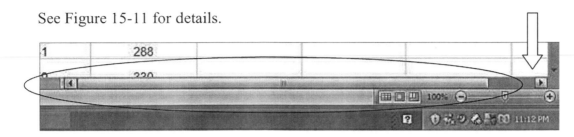

Figure 15-11

As you click the arrow the columns on the left will disappear one by one, all except column A. Column A will stay on the screen. This can be of great help if you have a lot of columns and you need to look at a column that is too far away to also be able to see the first column where all of the labels are located. It is easy to forget what is in each row when you can't see the labels in the first column.

To unfreeze a column you need to click on the Freeze Panes button a second time and choose Unfreeze Panes. The top choice is now the Unfreeze Panes command.

Unfreeze column A

You can always freeze the top row so that it will also remain visible. This is done exactly in the same way that a column is frozen.

Freeze the top row

Now you can use the scroll bar on the right side of the screen to see the rows below the visible area, and the top row with the labels will remain in view.

Unfreeze the top row

When nothing is frozen the top choice in the drop down menu will allow you to freeze panes according to the selected cell. You can click on a cell anywhere in the worksheet and then choose the top choice to freeze the panes above this point. Nothing above this line will move as you scroll down the worksheet.

Try it

When you are finished playing with your new toy:

Close the workbook without saving your changes

Chapter Fifteen   Review

Different views are available for different needs.

> The Normal View is the default view
> The Page Layout View will let you see how the page will look when it is printed.
> The Page Break View will allow you to see where the page breaks will be.
> The Full Screen View will remove everything from the screen except the worksheet.
> Custom Views will let you save a view with a different name.

The Show/Hide Group will let you decide if certain things, such as gridlines and headings are visible.

The Zoom command will let you view the screen as larger or smaller.

You can freeze columns or rows to keep them from moving, while you allow the rest of the worksheet to scroll.

Chapter Fifteen   Quiz

1) The Page Layout View will disable the Ribbon while you are using the view. **True or False**
2) When you are in the Page Break View, explain how you can move a page break.
3) Explain how to remove the Full Screen View.
4) Which group on the Views Tab has the command so you can choose not to see the Formula Bar?
5) The Zoom Slider at the bottom right of the screen is the only way to access the Zoom command. **True or False**
6) If you select column D and choose the Freeze Panes command, columns A, B, C, and D will remain frozen. **True or False**

LaVergne, TN USA
07 April 2010

178502LV00001B/26/P